T0134679

Innovation and Discovery in Russian Science and Engineering

Series Editors

Carlos Brebbia
Wessex Institute of Technology, Southampton, United Kingdom

Jerome J. Connor
Massachusetts Institute of Technology, Cambridge, MA, USA

More information about this series at http://www.springer.com/series/15790

Stavros Syngellakis • Carlos Brebbia
Editors

L. D. Gitelman • E. R. Magaril
Associate Editors

Challenges and Solutions in the Russian Energy Sector

 Springer

Editors
Stavros Syngellakis
Wessex Institute of Technology
Southampton, United Kingdom

Carlos Brebbia
Wessex Institute of Technology
Southampton, United Kingdom

Associate Editors
L. D. Gitelman
Ural Federal University
Yekaterinburg, Russia

E. R. Magaril
Ural Federal University
Yekaterinburg, Russia

ISSN 2520-8047   ISSN 2520-8055   (electronic)
Innovation and Discovery in Russian Science and Engineering
ISBN 978-3-030-09303-7   ISBN 978-3-319-75702-5   (eBook)
https://doi.org/10.1007/978-3-319-75702-5

Preface

This book was prepared through close cooperation between university researchers and practitioners. It is dedicated to the better understanding of the complex problems associated with the role of energy in today's world.

Energy ensures the functioning of almost all elements of modern society from large industrial and transport systems to computer devices and households.

It is a well-known rule that lower energy production in a country correlates to lower GDP (gross domestic product) and standards of living. Energy consumption in the world is also unevenly distributed. One billion persons consume 80% of energy, and the remaining 6 billion – only 20%. The available power in different countries differs by more than 50 times.

Modern economies can be supported only by countries which have managed to ensure abundant energy supplies. Economic growth is achieved through the development of the power industry.

The constant increase in population, the lack of energy resources and their uneven distribution, their impact on the environment, globalization and other factors are critical to achieving energy sustainability.

Judging by diverse forecasts, global energy consumption could double by mid-century. Growing tensions in the market of fossil fuels make it difficult to resolve this problem by increasing the use of traditional energy resources. The solution will require the use of nuclear energy and renewables. At the same time, there will be changes in the consumption of energy resources related to scientific and technical progress and to innovations in the world economy.

One of the major problems of modern society is to ensure free access of countries, businesses and individuals to energy and energy services. Considering the importance of the problem, the United Nations has put forward a global initiative in the form of the "Sustainable Energy for All". To implement it, a network of knowledge "The UN-Energy" and financing mechanisms of energy efficiency and renewable energy development have been created. Powerful intellectual capital, enormous material and financial resources are to be mobilized.

For cold-climate countries, the situation is exacerbated by the need to spend a significant portion of their energy for heating. For example, in Russia, the most urgent need for accelerated modernization of the energy sector is in the social and housing sector, which consumes about 80% of the country's power.

Problems of development of the energy sector in countries with transition economies relate to the increase in the ageing of the main equipment; insufficient levels of investment in the energy sector and their inefficient use; the absence of mechanisms to stimulate attraction of investments; lack of innovation in power engineering and electrical industries; large excessive losses of energy, especially in the heat supply systems; and ineffective tariff policy, low energy efficiency and irrational use of energy resources.

In the developed countries, the priority is to replace a significant part of the traditional non-renewable energy resources by new sources of energy, which requires joint efforts of scientists and the necessary financial resources.

Complex challenges associated with the acceleration of traditional energy consumption and associated pollution require an interdisciplinary approach to solve them. The approach ought to aim to cover all aspects, that is, innovative technologies to ensure energy efficiency and environmental safety of stationary and mobile power generating units and transport modes using hydrocarbons; introduction of alternative fuels and energy, and energy efficient technological processes. It is also necessary to take into account the social and political consequences of these tasks by implementing effective management systems based on the knowledge of engineering, economic and environmental factors. Extremely important is the education for managers providing them with the methodology to operate in an environment of accelerating changes, instability and high risks.

The chapters of this book attempt to discuss some of those problems and provide some guidelines regarding the future direction and new areas of research.

The editors would like to express their gratitude to all the authors for their contribution, and to the Editorial Board and other scientists who reviewed the material in each chapter and thus ensure the quality of this book.

The editors also gratefully acknowledge the support of Vice-Rector in Science of Ural Federal University V.V. Kruzhaev and Vice-Rector in Economic and Strategy Development D.G. Sandler.

Southampton, UK	Stavros Syngellakis
Southampton, UK	Carlos Brebbia
Yekaterinburg, Russia	L. D. Gitelman
Yekaterinburg, Russia	E. R. Magaril

Contents

Part I
Economic Problems of Energy Development

Assessing the Indicators of Municipalities' Energy Efficiency

V. V. Dobrodei

1 Introduction

The problem of spatial economic measurements at the level of regions and municipalities has recently received increased attention as territorial comparisons and assessments are necessary for the authorities as benchmarks in establishing strategies and programmes for social and economic development. Rating comparisons of social and economic development of the territories are the most common ones. Local ratings are used relatively less often, but that does not reduce their significance. In particular, these could include the assessments of the territories' environmental status and standards of living and of energy efficiency and energy security, etc. [1]. However, little attention is paid to improving the methodologies of calculating specific and aggregated rating indicators, to the correctness of assessment techniques, and to the identification of error intervals. All too often, approaches tend to simplify calculations and formal assessment of aggregated indicators at the expense of the detailed analysis of initial indicators and factors as sources of specific territorial issues.

The research which was conducted using statistical materials of Sverdlovsk Oblast identified the following information and methical problems of rating energy efficiency of municipal infrastructure.

Producing a ranking that is objective enough turns into a highly unconventional problem under these circumstances, considering the known complexity of the multi-parameter ordering task.

V. V. Dobrodei (✉)
Ural Federal University, Yekaterinburg, Russia

© Springer International Publishing AG, part of Springer Nature 2018
S. Syngellakis, C. Brebbia (eds.), *Challenges and Solutions in the Russian Energy Sector*, Innovation and Discovery in Russian Science and Engineering,
https://doi.org/10.1007/978-3-319-75702-5_1

2 Information and Methodological Aspects of Forming Energy Efficiency Ranking

It is known that comparing territories on the basis of individual indicators usually gives different orderings and rankings (for municipalities, in our case); therefore, the necessity of building aggregated ratings is obvious. The following information stages are important in substantiating the rankings: choosing initial indicators (among the available ones), building a system of indicators, and ensuring their comparability. As a rule, the initial array of indicators is currently insufficient to fully solve the task of territorial comparisons, especially at the level of municipalities. Therefore, the obtained aggregate ratings are, to some extent, of an individual nature.

As a result of analysing the statistical base of Russian municipalities, the following specific energy efficiency parameters were taken into account in the calculations:

P_1 is the electric power intensity of the companies operating within the territory of the municipality (hereinafter referred to as "electric power intensity of companies"), $P_1 = E/V_t$, kWh/thousand roubles, where E is the volume of electrical power sold to large- and medium-sized companies within the territory, kWh, and V_t is the turnover of large- and medium-sized companies operating within the territory of the municipality, thousand roubles.

P_2 is heating intensity of companies operating within the territory of the municipality (hereinafter referred to as "heating intensity of companies"), where $P_2 = T/V_t$, Gcal/thousand roubles, and T is the volume of heating supplied to large- and medium-sized companies within the territory, Gcal.

P_3 is the consumption of centrally supplied hot water per capital of the population living in housing with utilities within the territory of the municipality (hereinafter referred to as "HWS consumption per capita"), where $P_3 = V_{HWS}/P$, Gcal/person, and V_{HWS} is the volume of hot water sold to the population living within the territory, Gcal.

P_4 is the consumption of cold water per capita of the population living in homes with running water (hereinafter referred to as "water consumption per capita"), where $P_3 = V_{WP}/C$, cubic metre/person, and where V_{WP} is the volume of water sold to the population, thousand cubic metres.

P_5 is the consumption of thermal energy for the heating of a square metre of housing with utilities within the territory. It is defined as the ratio of the volume of thermal energy supplied to the population for heating and to the total surface area of residential premises equipped with central heating, multiplied by the corresponding climate factor: $P_5 = V_{he}/S$, Gcal/thousand square metres; V_{he} is the volume of thermal energy sold for heating to the population living within the territory, Gcal; and S is the total surface area of residential premises, equipped with centralized heat supply, thousand square metres.

P_6 is the ratio of leakages and unmetered water consumption to the total volume of water supplied into the network (hereinafter referred to as the specific weight of water losses), %. $P_6 = V_{vl}/V_v$; V_{vl} is leakage and unmetered water consumption, thousand cubic metres, and V_v is the volume of water supplied into the network, thousand cubic metres.

P_7 is the specific weight of thermal energy losses in the total volume of heat energy, supplied into the network (hereinafter referred to as the specific weight of heat energy losses), %. $P_7 = V_{hl}/V_h$; V_{hl} are heat losses, thousand Gcal; V_h is the volume of heat energy supplied into the network, thousand Gcal.

The use of the indicator "turnover of companies" when calculating the indicators P_1 and P_2 could be explained by the fact that the Russian statistics authorities do not calculate such indicator as "municipal product", and it is impossible to determine the energy intensity of products (works and services) produced in municipalities by means of gross value added. The calculation of individual national account indicators for municipalities in the Russian Federation was planned for the year 2016.

The choice of indicators P_3, P_4, P_5, P_6, and P_7 for the municipal energy efficiency scorecard is stipulated by their importance; the high relative energy intensity of the Russian economy is largely determined by the high level of actual losses in electricity and heat networks, as well as by the high degree of their deterioration and thermal insulation issues in buildings.

The pair correlation of the indicators used was checked to analyse their statistical relationship. Identification of strong dependence between the indicators would allow for revealing the situations in which the change of one indicator results in a substantial change in another, and it is necessary to decide whether it is feasible to combine them or not.

A moderate correlation is observed between the indicators "consumption of cold water per capita" and "HWS consumption per capita". It is obvious that both indicators depend on the non-observed factor – that of the proportion of people living in homes with utilities, i.e. there occurs technical data correlation (false correlation). A slight correlation is also observed between the indicators "power consumption of companies" and "consumption of thermal energy per square metre" and "ratio of water losses" and "consumption of thermal energy per square metre". As far as other indicators are concerned, weak or very weak correlation is observed between them. Therefore, the observed relationships between the indicators cannot be described as strong or very strong, which means it can be neglected when shifting the focus of calculations to the aggregate assessment of individual, but important aspects of energy consumption.

A database is compiled for each municipality that includes the following information: turnover of companies (total and per types of economic activities: mining; manufacturing; production and distribution of power energy, gas, and water; other activities, thousand roubles); electricity consumption by large- and medium-sized companies, thousand kWh; thermal energy consumption by large- and medium-sized companies, thousand Gcal; and number of people living in housing equipped with central hot water supply, persons, etc.

The author concentrated on the maximum possible integration between information and instrumental approaches and other documents on the social and economic development of the region when designing the energy efficiency ranking of

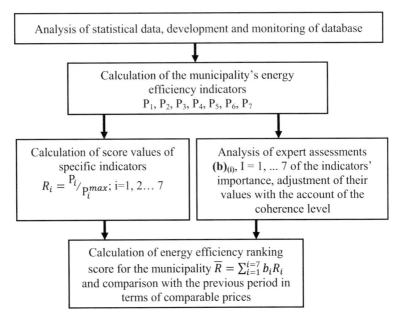

Fig. 1 Basic blocks of calculations in forming the municipality's efficiency ranking

municipalities. More specifically, they sought to combine them with the methodology of creating a ranking of social and economic development of the municipalities which provides for the assessment of production, financial, and social indicators per capita, but does not take into account the indicators reflecting the consumption of energy resources within the territory.

The sequence of calculations provides for comparative analysis of energy efficiency rankings for the reporting and base years (its basic steps are shown in Fig. 1).

The studied municipalities differ significantly, both in terms of population and their economic potential, so in order to improve their homogeneity, the whole set of them was divided into three groups depending on the value of the "turnover of the companies" indicator. The analysis of the ranking calculation results showed that such a grouping is reasonable. Indeed, the objective differences in the turnover encompass not only the peculiarities of the municipalities' weather conditions but also the structure of their economy, its efficiency, and other factors that remain unchanged when calculating the aggregate indicators. These differences are of a long-term nature, so the factor of "turnover" may be used as an indication of territorial differentiation.

The assessment of specific rankings is incorporated into the general scheme of calculations, their integration into the system of aggregate ratings being differentiated as per groups of municipalities, thus meeting the conditions of relative homogeneity. The group boundaries are not determined expertly, which reduces the level of assessment subjectivity. The arguments in favour of the feasibility of such operations are the following: possible inaccuracies in calculating the indicators are smoothed over within the interval boundaries, and there is a chance for a regional policy to appear that would be differentiated in relation to different groups of municipalities.

Expert weighing of indicators and non-recognition of them as having equal significance introduce some subjectivity into the resulting aggregate indicators. However, the logical foundations of this process are undeniable because the very presence of a number of individual indicators determines their comparative significance. In our case, transition to the creation of an ordered system of group rankings enables the formation of a consistent unified system of municipalities' ranking, provided that a correct choice is made of the standardized weights of individual groups, i.e. the scorecard for the assessment of municipalities is hierarchically ordered.

The methodology used the procedure for calculating average weight factors of indicators within the ranges of 95% of their consistency with expert opinions. The questionnaire for the survey provides for flexible consideration of expert opinions in the form of an interval indicating upper and lower boundaries (not necessarily integral values) within the range from 0 to 10. The standardized average weights in the range [0;1] are calculated according to the results of expert interval assessment. Then, the scope of expert assessments' variation is determined for each indicator. Due to the fact that the sample size is small (15 experts), the random number generator was used in the range of variation in order to obtain confidence intervals of weight factors with a predetermined level of significance of 0.95. The initial distribution of the average weight factors most often turns out to be close to normal. The obtained mathematical expectations of weights are adjusted with the help of the least square method provided that the sum of the final coefficients is equal to one. As a result, the final weights of the indicators have a two-component structure – adjusted average value (very close to the mathematical expectation) and confidence intervals (variation without tails). On the basis of these data, rank orderings can be made, and it is possible to assess appropriate intervals of aggregated rating indicators. Table 1 shows the intervals of weight factors based on the materials for Sverdlovsk Oblast for the year 2013.

Table 1 Intervals of standardized weight factors of energy efficiency indicators

	P_1	P_2	P_3	P_4	P_5	P_6	P_7
Minimum of average expert assessments	0.019	0.019	0.035	0.044	0.109	0.071	0.099
Maximum of average expert assessments	0.214	0.200	0.192	0.163	0.244	0.227	0.227
Average value of expert assessments	0.119	0.119	0.137	0.128	0.181	0.146	0.170
Standard deviation	0.054	0.048	0.040	0.030	0.033	0.050	0.039
Mean value	0.118	0.109	0.136	0.103	0.176	0.145	0.169
Confidence interval	**0.106**	**0.098**	**0.127**	**0.096**	**0.168**	**0.134**	**0.160**
	0.129	**0.119**	**0.145**	**0.110**	**0.184**	**0.156**	**0.177**

Final weight factors: $b_1 = 0.127$; $b_2 = 0.119$; $b_3 = 0.135$; $b_4 = 0.110$; $b_5 = 0.184$; $b_6 = 0.156$; $b_7 = 0.170$.

In each group there are so-called leaders and outsiders as per the basic values (without weighting), and specific causes of such positioning are clarified. For this reason quartiles for each indicator are calculated, and the composition of the municipality is analysed. The municipalities that stand out from the total sample (emissions) are identified by means of scattergram analysis. The outliers might be caused by an atypical but quite real situation as well as by incorrect basic information. It is advisable to conduct statistical analysis of the distribution of indicators within each group, which allows for determining the mean value of the indicator and its confidence interval with the specified level of reliability, as well as the characteristics of how close the distribution type is to the normal one.

As it had been expected, the distribution tails have different forms, but most distributions are close to symmetric type, the mean one being close to the midpoint, even though they are not strictly normal. It is advisable to conduct similar operations also for ranking (aggregate weighted) indicators for each group of municipalities using the tools of descriptive statistics, histogram analysis, calculation of the energy efficiency quartiles, and analysis of respective composition of the municipality.

Rescaling nominal indicators of the current year to the base year was carried out by means of respective turnover recalculation for each municipality, taking into account price indices for each type of economic activity C (mining), D (manufacturing), E (production and distribution of electricity, gas, and water), and an aggregate index of consumer prices for goods and services (other activities) [2]. The analysis of the results of comparing selected groups on the basis of cumulative curves should be complemented by the structural analysis of the obtained shiftings. These methods are illustrated in Figs. 2 and 3 for the municipalities in the first group (part of a group with relatively high business turnover).

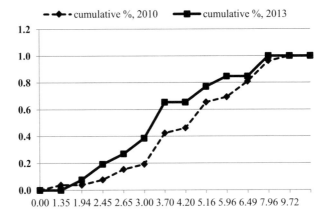

Fig. 2 Curves of the cumulative frequencies of rank energy efficiency values of municipalities of Group 1 for the years 2010 and 2013 in comparable conditions

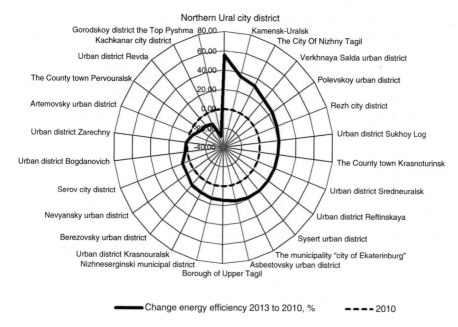

Fig. 3 Changing energy efficiency of the municipalities in Group 1

The results of the analysis confirm the fact that on the whole the region witnesses a decrease in the final ranking values and, therefore, a large part of the municipalities in the region was characterized by an increase in energy efficiency in the year 2013 as compared to the year 2010. In the year 2013 the improvements in energy efficiency occurred in 39 municipalities out of the 54 studied municipalities. Specific indicators providing for the observed effect and factors impeding the growth of energy efficiency were identified.

It is known that the effects of pseudo-compensation can occur in the scope of the aggregate rankings when low values of some indicators are compensated by high assessments of the others, thus determining a higher aggregate score. The resulting illusion of prosperity excludes such municipalities from the list of those in need of balancing.

3   Conclusions

The author believes that the methodology should be further developed with the purpose of making the analysis more comprehensive, and the statistical base of the municipalities should be supplemented with other indicators that take into account the requirements of federal and regional legislation in the field of energy conserva-tion and energy efficiency of municipalities. It should be noted that the indicators, on which the scorecard is based, focus more on the characteristic of the current levels of energy efficiency in municipalities. Under the conditions of limited information, the assessments are approximate despite the detailed analysis of the municipalities' economic specialization. The credibility of assessments can be improved if more elaborate indicators of socioeconomic development of the municipalities in the region are obtained. To assess energy efficiency potential, it is necessary to take into account the indicators that stimulate the efficiency of energy consumption. The effectiveness of such tools is confirmed by international experience [3].

References

1. Bashmakov, A., Myshak, A.: Contribution of regions to the dynamics of the energy intensity of GDP in Russia [in Russian]. Energy Conserv. **8**, 12–18 (2013)
2. Socio-economic Situation of Sverdlovsk Oblast. January-December 2013: Analytical Review [in Russian]. Territorial body of the Federal State Statistics Service of Sverdlovsk Oblast, Ekaterinburg (2014)
3. Bashmakov, I.A., Bashmakov, V.I.: Comparing Russian Policy on Improving Energy Efficiency with Measures Being Taken in Developed Countries. Centre for the Effective Energy Use (CENEF), Moscow (2012)

Methods of Evaluation of the Market Power Level on the Wholesale Electricity Market

A. Trachuk and D. G. Sandler

1 Introduction

Over the last few years, much attention has been paid to theoretical research and practical work toward the implementation of competitive relations in the industries traditionally referred to as natural monopolies [1–3].

In most countries that have experienced restructuring of the electric power industry, it was planned that competition would have a positive impact on pricing and generate signals for industry development, which was originally built on the monopoly principles. The study is an attempt to answer the question: how to assess the level of competition in the reformed electricity market or, in other words, the market power level. For this purpose, we examine the wholesale power market.

The authors believe that, apart from determining approaches to the creation of a competitive environment in naturally monopolistic industries, it is important to propose methods of evaluating the competitive level as well. Books and articles propose different approaches to measuring market powers [4, 5]. At the same time, authors focus their attention primarily on competition among sellers.

For instance, one of widely used approaches is an approach developed within the framework of the economic theory and based on an analysis of concentration and monopolistic power. The nature of such approaches and the interrelation between them should be appropriate to consider in the case study of Lerner and Herfindahl–Hirschman indexes [6].

A. Trachuk (✉)
FSUE Goznak, Financial University Under the Government of the Russian Federation, Moscow, Russia

D. G. Sandler
Ural Federal University, Yekaterinburg, Russia

© Springer International Publishing AG, part of Springer Nature 2018   11
S. Syngellakis, C. Brebbia (eds.), *Challenges and Solutions in the Russian Energy Sector*, Innovation and Discovery in Russian Science and Engineering, https://doi.org/10.1007/978-3-319-75702-5_2

2 Research Methodology

The monopolistic power level of a firm on the market could be evaluated by determining a difference between the price and marginal costs. The more the price asked by the firm deviates from marginal costs, the higher market power the firm seizes and hence the higher level of the market monopolization is. The Lerner index is defined as a difference in prices of competitive and noncompetitive markets with respect to a noncompetitive price:

$$L = \frac{P_m - P_c}{P_c}, \tag{1}$$

where L is Lerner index, P_m noncompetitive market price, and P_c competitive market price.

In the long run the price in case of perfect competition is equal to marginal costs, yet it is difficult to determine actual marginal costs, so they can be considered to be approximately equal to average marginal costs. The Lerner index can be defined as a difference between the price and average variable costs with respect to the price, and, given that the price and marginal costs are interrelated through elasticity of demand for the price and that on the competitive market elasticity of demand for the price is endless for an individual firm, the price is equal to marginal costs, and the Lerner index can be presented as follows:

$$L = \frac{P - AVC}{p} = -\frac{1}{E_d}, \tag{2}$$

where P – price of a product of this firm (on this market); AVC – average costs of the firm (this firm); E_d – elasticity of demand.

The Lerner coefficient ranges from zero (on a market with perfect competition) to one. The higher the index is, the higher the monopolistic power and the greater the gap between the market and an ideal condition of perfect competition. The problem of applying the Lerner index in practice is preconditioned by the fact that demand for services of natural monopolies in the short run is practically nonelastic with respect to the price.

In practice, the Herfindahl–Hirschman index (HHI) [7, 8] is mostly used to evaluate market competition, which is determined as a sum of the squares of shares of all the firms operating on the market:

$$HHI = \sum_{i=1}^{n} Y_i^2, \tag{3}$$

where Y – market share of a firm operating on the market.

The HHI ranges from 0 (in case of ideal competition when there is an endless number of sellers on the market and each of them controls an insignificant market share) to 1, which corresponds to a monopoly. The main advantage of the index is a capacity of giving information about the market power of certain companies and assessing the level of concentration in the industry.

It is worth noting that, proposed as a method of market concentration evaluation, this indicator has been transformed into the market power indicator. Theoretically, this is associated with a relation between the Lerner index and market concentration in case of an oligopoly.

Let us assume that such market is described by the Cournot model. The Cournot model is based on an assumption that the firm establishing the scope of sales considers that the scope of sales of other firms is invariable. In this case, for firms interacting according to Cournot, the Lerner indicator for the firm will be directly dependent on the firm's share market (ratio of the scope of sales on the market to the industry-specific scope of sales) and inversely dependent on the demand elasticity indicator:

$$L = -\frac{Y_i}{E_d}. \tag{4}$$

The average Lerner index for the industry (where shares of firms on the market serve as weights) is calculated according to the formula below:

$$L = -\frac{HHI}{E_d}, \tag{5}$$

where HHI is Herfindahl–Hirschman concentration index.

Interrelation between the concentration indicator (Herfindahl–Hirschman index) and the monopolistic power indicator (Lerner index) is widely used in empirical research despite obvious theoretical limitations associated with the need to assess the elasticity of demand and a condition that competition according to Cournot is present.

It is sometimes indicated [9] that the HHI is not fully suitable for analyzing the electricity market as it fails to take into account practical lack of elasticity of demand, practically total dependence of certain suppliers on actions of other suppliers (competition not according to the Cournot model), existence of long-term liabilities and forward contracts, and difficulties in determining the market boundaries. It should be noted that the first two factors in general increase the market power and, therefore, even an HHI below 0.18 may not guarantee lack of the supplier's market power.

However, according to authors, application of the Herfindahl–Hirschman index is completely justified in case of the reformed RF power industry characterized by quite a complicated network structure with multiple network and system limitations.

For this reason, one more aspect, i.e., market boundaries, should be studied to prove a possibility of using this approach.

System limitations divide the uniform price zone market into local segments: the zone of free transmission (ZFT) of electrical energy and hubs. In fact, the single competitive selection is split into several smaller tenders limited primarily to the frameworks of free power transfer zones. Therefore, competition should be determined not only according to the market in general or price zones, but within the frameworks of free power transfer zones as well.

Furthermore, competition in general on the market or price zones could be evaluated through set-theoretical unions of results for ZFT (volumes within the price zone are aggregate volumes in certain ZFT).

Taking into account the aforesaid circumstances and the fact that the share of direct and forward contracts between suppliers and buyers in the electricity whole-sale market is relatively low, it is appropriate to apply the *HHI* to assess the market power of the wholesale market suppliers.

3 Testing of the Proposed Approach

The presented approach could be demonstrated by using the case of the ZFT Tyumen. The following plants are located in specific zone (Table 1).

Given the calculated shares of suppliers, the *HHI* of this free power transfer zone is equal to $(15.7\%)2 + (15.7\%)2 + (25.3\%)2 + (43.3\%)2 = 0.220$. Total transfers from ZFT Tyumen are equal to 1.143 MW, i.e., total output in the zone in the worst case (with highest possible transfers from ZFT) amounts to $11.359 - 1.143 = 10.216$ MW. The peak demand across ZFT totals 9.967 MW. Therefore, at the peak time only 249 MW of "unclaimed capacity" are not loaded. Any entity's capacity is much higher than the amount of unclaimed capacity. Therefore, each entity in the zone in the worst case (peak consumption and maximum transfers from the zone) has an exclusive status.

For this zone of free transmission of electrical energy the maximum incoming transfers amount to 1.790 MW. Therefore, in the best case for transfers (all transfers are directed to the zone), unclaimed capacity is equal to $11.359 + 1.790\ 9.967 = 3.182$ MW. Even in the optimal case for transfers E.ON and OGK-2 have an exclusive status.

Table 1 Characteristics of different power plants

Power plant	Company	Major shareholder (owner)	Available capacity, MW	Share in the ZFT
Tyumen CHPP-1	JSC Fortum	Fortum (Finland)	620	4.8%
Tyumen CHPP-2	JSC Fortum	Fortum (Finland)	755	5.8%
Tobolsk CHPP	JSC Fortum	Fortum (Finland)	665	5.1%
Total for TGK-10			**2.040**	**15.7%**
Nizhnevartovskaya GRES	CJSC Nizhnevartovskaya GRES	JSC inter RAO	2.013	15.7%
Surgut GRES-1	OGK-2	JSC Gazprom-energoholding	3.268	25.3%
Surgut GRES -2	E.ON Russia	E-on (Germany)	5.597	43.3%
ZFT total			12.918	100%

It is also interesting to study the extent to which the proposed approach can be used for evaluating the competitive situation for demand. Taking into account the fact that the retail market players are only partly involved in changes that happened on the electricity market, competition between buyers of the wholesale electricity and capacity market (WECM) should be considered.

It should be noted that creation of conditions for competition was one of the most important objectives of reforming the electricity market in most of developed and developing countries. However, a decade after the reform was launched, one could state that the said objective has not been achieved [10]. If the reasons of domination of generators in certain free power transfer zones are mostly determined by the deficit of generating capacities and (or) specifics of the power network topology, competition between consumers is lacking for somewhat different reasons.

In this section of the chapter, we consider the segments of the Russian power market where competition between buyers is possible, i.e., nonregulated trade in electricity (Day-Ahead Market (DAM), balancing market) and conclusion of free bilateral electricity and/or capacity purchase and sale agreements.

However, evaluation of a competitive condition of market sectors should be preceded by a description of the structure of WECM buyers as specifics of partic-ipating in the market of buyers of different types form prerequisites for a possible competition pattern.

The subjective structure of the wholesale electricity (power) market of Russia changes quite often, but the shares of categories of buyers change only slightly. Last resort suppliers account for about 50% of the total number of WECM buyers (according to the list of buyers in the WECM entities available on the official website of NP Market Council, www.np-sr.ru) and major end consumers for about 15%, and power-selling companies without a LRS status account for 35%. When evaluating the shares of types of buyers in terms of the volume of consumption, the LRS share drastically declines: for instance, in the second price zone (Siberia), the share of last resort suppliers slightly surpasses 50% in terms of the number and totals about 83% in terms of the volume.

This LRS domination can be explained by the fact that in most cases they are companies split-off upon restructure of regional power JSCs supplying electricity to the entire area, and redistribution of their volumes to other power-selling entities, including the cases of large consumers entering the wholesale market, is slow.

It should be borne in mind that last resort suppliers must submit price-accepting applications to the DAM. However, other groups of consumers authorized to submit price applications almost do not exercise this right (the DAM demand curve in most hours is elastic only in a small section, 1–2% of planned consumption).

Sales organizations (LRSs and those that are not last resort suppliers) buy electricity and capacity in the wholesale market to sell them on the retail market. Major end consumers entering the WECM buy electricity (capacity) for their own consumption needs. This segment of WECM buyers is represented by major indus-trial plants, normally energy-intensive ones. This structure of end consumers of electricity acquiring it on the WECM can be explained by the fact that the requirements for operation on the wholesale market are quite complicated and expensive for many consumers.

Economic benefit as a result of entering the wholesale market for the end consumer consists, first and foremost, in declining expenses of buying electricity as a result of saving on intermediaries, i.e., as a result of saving on the sales premium of its retailer. Therefore, participation of a major consumer in the trade on the wholesale market is economically justified in case where sales premium economy is higher than expenses of bringing the record and communication system in line with requirements for them.

In certain cases, before entering the wholesale market, large consumers of electricity possessed their own sales from which electricity was purchased, but an analysis of the financial result of performance of such energy-selling organizations showed that for consumers it is cheaper to buy electricity directly on the WECM, so they decided to enter the wholesale market.

For some large industrial companies, a decision to buy electricity on the wholesale market was purely a political and (or) image-building move.

It should be noted that even companies within the same industry with comparable energy consumption and similar financial standing reach different decisions on entering the wholesale market. This can be explained by a difference in tariffs applicable in different regions of Russia, a difference in relations of management of an industrial company and an electricity-selling company and the like, and in certain cases the above noneconomic causes of entrance in the wholesale market by large consumers.

As in the current DAM model with price applications of buyers, the main parameter determining the competitive condition of the market is its concentration (volume shares of players) assessed, in particular, with the help of Herfindahl–Hirschman indexes [11].

Volumes of consumptions by players change every hour; therefore, the average purchase volumes in statistically important periods can be used to calculate the shares of buyers in electricity markets. However, the peak, not average, consumption of buyers in the wholesale market is used to trade power; therefore, we believe that shares of consumers in all sectors of the WECM should be calculated based on peaks of participants. On average, shares of participants calculated according to peak consumption, slightly differ from shares calculated according to average statistical volumes, which makes it possible to consider calculation of market concentrations based on peaks of consumers to accurately reflect the competitive condition of both capacity markets (characterized for consumers by their peaks) and energy markets (measured by average or total (integral) consumption over a period).

Let us analyze the market shares of buyers for evaluating competition among WECM buyers by free power transfer zones (ZFT) and hubs.

As was mentioned, records of network and system limitations are important for analyzing the market power. These limitations divide the whole market of the target zone into local segments: free power transfer zones (ZFT) and hubs.

Let us consider the results of this calculation of the above ZFT Tyumen (Table 2).

Table 2 Results of calculations

ZFT Tyumen

Company	Group of companies	Peak of consumption, MW	Share in the ZFT
JSC Tyumen energy-selling company	Gazprom	3,639,465	62.4%
Surgut GRES-1	Gazprom	84,051	1.4%
Total for Gazprom Group		**3,723,516**	**63.8%**
Nizhnevartovsk GRES	Inter RAO	23,209	0.4%
LLC RN-Energo	Rosneft	359,614	6.2%
Total for groups controlled by the state		**4,106,339**	**70.4%**
JSC Omsk energy-selling company	Energostream	1459	0.0%
JSC Tomsk energy-selling company	Energostream	137,812	2.4%
Total for Energostream Group		**139,271**	**2.4%**
Plants of JSC Fortum	Fortum	83,741	1.4%
JSC Tyumenenergosbyt	Energy sales holding	302,740	5.2%
LLC Nizhnevartovsk energy-selling company	NESKO	65,752	1.1%
LLC Noyabrsk steam-gas power plant	Intertechelectro	304	0.0%
CJSC EESnK	TNK-BP	784,683	13.4%
LLC Rusenergoresurs	ESN	260,300	4.5%
Municipal electrical networks of Khanty-Mansiysk	ChMGES	33,906	0.6%
Surgut GRES-2	E-on	59,744	1.0%
ZFT total		**5,836,781**	**100%**

HHI total $= 63.8\%^2 + 0.4\%^2 + 6.2\%^2 + 2.4\%^2 + 1.4\%^2 + 5.2\%^2 + 1.1\%^2 + 0\%^2 + 13.4\%^2 + 4.5\%^2 + 0.6\%^2 + 1\%^2 = 0.435$

The level of market concentration in this ZFT is quite high as buyers in these zones are normally represented by one or two major sales companies holding large shares of the market and separated from local energy JSCs as a result of their restructure and supplying power to the relevant RF territories and also a number of small consumers in the wholesale market (small sales and power plants consuming electricity for their own economic purposes).

4   Conclusion

In the study, the authors justified the application of the Herfindahl–Hirschman index for the purpose of practical evaluation of the market power level on the wholesale electricity and capacity market. An analysis of actual information about the competition levels both among sellers and buyers in one of the free power transfer zones shows that price offers among sellers on the wholesale electricity and capacity market are currently formed primarily by suppliers that may ensure supplies taking into account the existing network and system limitations, and a small number of last resort suppliers is predominant among buyers. The oligopoly of suppliers and a similar market condition among WECM buyers form a market structure of oligonomy, which emphasizes again the need for state regulation in this area, however, by taking into account new realities: the oligopoly type market.

References

1. Green, R., Newbery, D.: Competition in the British electricity spot market. J. Polit. Econ. **100**(5), 929–953 (1992)
2. Joskow, P.: Restructuring, competition and regulatory reform in the U.S. electricity sector. J. Econ. Perspect. **11**(3), 119–138 (1997)
3. Joskow, P.: Lessons learned from electricity market liberalization. Energy Journal. **29**(SPEC. ISS), 9–42 (2008)
4. Wolfram, C.: Measuring duopoly power in the British electricity spot market. Am. Econ. Rev. **89**(4), 805–826 (1999)
5. Bushnell, J., Mansur, E., Saravia, C.: Vertical arrangements, market structure, and competition: An analysis of restructured US electricity markets. Am. Econ. Rev. **98**(1), 237–266 (2008)
6. Trachuk, A.V.: Evaluation of a competitive environment on the wholesale electricity market [in Russian]. Econ. Sci. **66**, 124–130 (2010)
7. Wang, P., Xiao, Y., Ding, Y.: Nodal market power assessment in electricity markets. IEEE Trans. Pow. Syst. **19**(3), 1373–1379 (2004)
8. James, B.: Oligopoly equilibria in electricity contract markets. J. Regul. Econ. **32**(3), 225–245 (2007)
9. Stoff, S.: Power Systems Economics: Designing Markets for Electricity. IEEE/Wiley, New York (2002)
10. Gitelman, L.D., Ratnikov, B.E.: Reform of the power industry: Evaluation of efficiency and course correction, [in Russian]. Energy Market. **1**, 10–14 (2009)
11. McConnell, B.: Economics: Principles, Problems, and Politics. INFRA-M, Moscow (1999)

Methodological Tools for Energy Efficiency Assessment of Industrial Systems in a Context of Transition to a "Green Economy"

V. Krivorotov, A. Kalina, A. Savelyeva, and S. Erypalov

1 Introduction

At present, industrial systems associated with major integrated structures acting as the locomotives of development of socioeconomic systems are becoming a core object of activity of economic systems. The authors adhere to the following definition of the term "industrial system": "an industrial system is an integrated structure consisting of manufacturing facilities of the main manufacturing cycle which are consolidated within a single chain of manufacturing operations from design and production to distribution and after-sales service, as well servicing companies which are engaged in joint work and enjoy synergistic benefits" [1].

As for energy efficiency of contemporary industrial systems, it is worth mentioning that many of them are characterized by high consumption of energy resources and high per-unit capacity of energy consuming facilities.

In this context, energy efficiency monitoring both at the level of the national economy and that of individual industrial systems is of great importance.

2 Analysis of Contemporary Approaches to Energy Efficiency Assessment

In the most general sense, energy efficiency in manufacturing is usually measured in two ways: the ratio of energy consumption to the yield of the product of technological processes (output of goods, services, work or other kinds of energy) and rational

V. Krivorotov (✉) · A. Kalina · A. Savelyeva · S. Erypalov
Department of Economics of Industrial and Energy Systems, Ural Federal University,
Yekaterinburg, Russia

© Springer International Publishing AG, part of Springer Nature 2018
S. Syngellakis, C. Brebbia (eds.), *Challenges and Solutions in the Russian Energy Sector*, Innovation and Discovery in Russian Science and Engineering,
https://doi.org/10.1007/978-3-319-75702-5_3

(or efficient) use of energy, i.e., the use of energy as much as needed, in the right way and at the right time.

In both cases the assessment of energy efficiency is based on a number of indicators, with their composition being specific in each individual situation. The analysis of numerous works by both domestic and foreign scholars makes it possible to single out two major approaches to selecting indicators and subsequent energy efficiency assessment of industrial systems and processes.

The first approach divides energy efficiency indicators into economic (cost-based), technical and economic (physical), and indicators showing the amount of energy-efficient technologies used. The World Energy Council uses a method based on this approach for energy efficiency assessment [2–4]. Another approach divides indicators of energy efficiency based on the types of activity. Among those using the abovementioned method are the Asia-Pacific Research Center [5], the project on indicators of the International Energy Agency [6–8], the project of the French Environment and Energy Agency ADEME [9], the technical service on strategy and energy efficiency of the World Energy Council [10, 11], and the Lawrence Berkeley National Laboratory [12].

Each of the abovementioned approaches has its advantages and disadvantages that have been discussed by experts. Generally, both approaches are applied together: at a more aggregated level economic indicators are used while when conducting a study of the industry technical indicators are applied, and at the most disaggregated level only technical indicators tend to dominate.

It is noteworthy that it is common practice worldwide to use the indicator of final energy consumption used for product manufacturing for determining the product's energy intensity [12]. But energy efficiency assessment of an industrial system cannot be limited to measuring its energy intensity; there should be a system of indicators to fully reflect the energy component of manufacturing.

3 Methodological Approach to Energy Efficiency Assessment of Industrial Systems

Based on the analysis of works compiled by Russian and non-Russian scholars, the authors propose a system of indicators to assess energy efficiency which operates at three levels of an industrial system: level of the industrial system as a whole, level of specific kinds of goods produced by the industrial system, and level of technological processes of manufacturing.

Energy efficiency can be measured for the industrial system as a whole or for each manufacturing process. Such kind of analysis requires more data, but it will provide the most accurate information. It is necessary to reveal consistent patterns in the correlation between economic growth and an increase in energy consumption. Nowadays developed countries focus on the strategy of the "green economy" when working out strategies of long-term development.

For instance, Anufriyev et al. [13] regard greenhouse gas and pollutant emission reduction as the main "green economy" indicator. The OECD singles out four groups of indicators describing the social and economic situation of a country and the characteristics of "green economy" growth [14]. When considering the energy efficiency indicators from the point of view of a "green economy," we take a closer look at the indicators of the ODYSSEE database dealing with energy efficiency [15].

In accordance with the IEA, CO_2 emission evaluation may be based on the following indicators: CO_2 emission/cumulative energy resources consumption (tCO_2/TJ), CO_2 emission/revenue ($kgCO_2$/rouble), and CO_2 emission/kW [16].

The IEA Energy Sector Carbon Intensity Index (ESCII) tracks how many tonnes of CO_2 are emitted for each unit of energy supplied. Since 1990, however, the ESCII has remained essentially static, changing by less than 1%, despite the important climate policy commitments at the 1992 Rio Conference and under the 1997 Kyoto Protocol as well as the boom in renewable technologies over the last decade. The ESCII shows only one side of the decarbonization challenge: the world must slow the growth of energy demand as well as make its energy supply cleaner [17].

HSB Solomon Associates successfully base GHG metrics primarily on the Energy Intensity Index (EII) – an industry standard for nearly 30 years [18].

The methodology proposes the following indicators to characterize transition to the "green economy" concept: ratio of CO_2 emission to primary consumption of energy resources (gram of CO_2/gram of reference fuel), ratio of CO_2 emission to gross revenue in comparable prices (gCO_2/rouble), and ratio of CO_2 emission to primary electricity consumption (gCO_2/kilowatt-hour).

As we have mentioned before and as can be seen from Table 1, when analyzing energy efficiency of an industrial system, the following levels are examined: the level of the industrial system as a whole (E_1), the level of specific goods produced by the industrial system (E_2), and the level of technological processes of manufacturing. Level E_1 is the most aggregate one and deals with consolidated data on the industrial system without going into detail. Level E_1 is the level of pooled analysis and is aimed at revealing the main energy efficiency problems of the industrial system. Level E_2 is the level of energy efficiency assessment of the industrial system's operation based on certain kinds of products the system turns out. Level E_3 is the level of a more in-depth energy analysis of the industrial system's operation. It is extremely difficult, sometimes impossible, to obtain data for energy assessment at this level (for instance, it may require the installation of complex and costly measurement equipment).

At levels E_2 and E_3, assessments built upon some indicators are not always equivalent, i.e., have different significance in the shaping of the energy efficiency assessment. That is why measuring the significance of specific groups of indicators should be another step when assessing energy efficiency. We suggest using the analytic hierarchy process with the help of a scale as in Ref. [19].

As for level E_1, all assessments of different groups are regarded as equivalent (equal ranking). That is why we can propose the following equation for determining the resultant energy efficiency assessment at k-level of evaluation:

Table 1 Groups of energy efficiency indicators of industrial systems

Indicators	Unit of measurement
1. Level of industrial system (E_1)	
Group 1.1. Efficiency of energy-consuming systems and energy saving	
1. Energy intensity of gross revenue in terms of primary energy consumption in comparable prices	Gram of reference fuel/ RUB
2. Electricity intensity of gross revenue in terms of primary energy consumption of revenue	Kilowatt-hour/thousand of roubles
3. Energy efficiency of gross revenue in terms of final (secondary) energy consumption in comparable prices	Gram of reference fuel / RUB
4. Electricity intensity of gross revenue in terms of final (secondary) energy consumption of revenue	Kilowatt-hour/thousand of roubles
5. Electricity losses in electric networks of industrial system	%
6. Heat losses in heating networks	%
Group1.2. Economic efficiency of energy consumption and efficiency of fixed asset use	
1. Share of fuel and energy costs in gross revenue	%
2. Coefficient of advance growth of electricity consumption in relation to production growth	Relative unit (%)
3. Coefficient of advance growth of primary energy consumption in relation to production growth	Relative unit (%)
4. Energy intensity of fixed assets	Gram of reference fuel/ RUB
5. Electricity intensity of fixed assets	kWh/thousand of roubles
Group 1.3. Eco-efficiency of fuel and energy resources consumption	
1. Ratio of CO_2 emission to primary energy consumption	gCO_2/ gram of reference fuel
2. Ratio of CO_2 to gross revenue in comparable prices	gCO_2/RUB
3. Ratio of CO_2 emission to primary electricity consumption	gCO_2/ kWh

$$E_k = \sum_{n=1}^{N} X_{kn} \cdot a_n; \sum_{n=1}^{N} a_n = 1, \qquad (1)$$

where X_{kn} is the resultant energy efficiency assessment of the industrial system at k-level for n-group; N is the number of groups of indicators that energy efficiency assessment at k-level is based on; a_n is the factor of the significance of n-group in the shaping of integral energy efficiency assessment at k-level.

All indicators have different and disparate units of measurement. In order to obtain assessments for groups of indicators X_{kn}, and the integral energy efficiency assessment at each level E_k, one has to resort to the procedure of standardization for individual indicators, i.e., convert them into a comparable dimensionless form. The authors suggest the following approach:

$$\alpha_{kns}^{norm} = \frac{\alpha_{kns}}{\alpha_{kns,\,basic}}, \tag{2}$$

where α_{kns}^{norm} is the standardized value of indicator s, included in group n at k-level of energy efficiency assessment; α_{kns} is the current (actual) value of indicator s, included in group n at k-level of energy efficiency assessment; $\alpha_{kns,\,basic}$ is the basic value of indicator s, included in group n at k-level of energy efficiency assessment.

For basic values one usually takes values of similar indicators of the main competitors or target values determined by the development strategy of the industrial system for a future period. In order to do so, one may use the geometric mean of indicators comprising the group:

$$X_{kn} = \sqrt[Z]{\prod_{s=1}^{Z} \alpha_{kns}^{norm}}, \tag{3}$$

where Z is the number of indicators comprising group n at k-level.

A similar procedure is used to determine an integral indicator at level E_1, unlike levels E_2 and E_3 where integral indicators are determined by Eq. (1).

When assessing the energy efficiency at three levels it is necessary to obtain the total resultant assessment E_{total} in addition to assessments at each level E_k. This combines assessments at all levels and is the most objective integral criterion of the system's energy efficiency. Assessments at levels E_2 and E_3 are not always available. In this case, E_{total} may be calculated with the following Eq. (4):

$$E_{total} = \sqrt[M]{\prod_{m=1}^{M} E_1^m \cdot K_2 \cdot K_3}, \tag{4}$$

where E_1^m is the integral assessment of energy efficiency at the first level obtained for a production unit of industrial system m; M is the number of main facilities of the industrial system; K_2 is the coefficient which takes into account energy efficiency assessments obtained for level E_2; K_3 is the coefficient which takes into account energy efficiency assessments obtained for level E_3.

For their part, values K_2 and K_3 are determined in the following way:

$$K_2 = \sum_{i=1}^{R} E_{2i} \cdot b_i; \quad \sum_{i=1}^{R} b_i = 1, \tag{5}$$

where E_{2i} is the resultant energy efficiency assessment obtained for i-product produced at the manufacturing complex; R is the number of main types of goods produced by the industrial system; b_i is the share of i-product in gross revenue.

$$K_3 = \sum_{j=1}^{N} E_{3j} \cdot c_j; \sum_{j=1}^{N} c_j = 1, \qquad (6)$$

where E_{3j} is the resultant energy efficiency assessment obtained for j-technological process for production of goods at the manufacturing complex; N is the number of main technological processes used by the industrial system; c_j reflects the significance (weight) of j-technological process in the operation of the industrial system. Values of c_j are determined by expert judgment.

It should be mentioned that the proposed methodological approach is all-purposed, and given adaptation of indicator system, it can be applied to industrial systems of different activities. The comparative analyses can be performed for industrial systems in different world regions. But nevertheless these analyses should be carried out inside one industrial activity.

4   Validation of a Methodological Approach Through Application to Assessment of Basic Energy Efficiency Indicators at the Ural Mountain-Metal Company

The proposed methodological approach was tested in a study of Russia's biggest holding company, the Ural Mountain-Metal Company (UMMC). The UMMC consists of almost 50 companies with an aggregate annual turnover of several billion dollars. The core of the company consists of a complete copper production cycle. We selected companies, engaged in copper ore mining, processing and crude, and cathode copper production: JSC Svyatogor, JSC Ural Copper Plant, and JSC Uralelektromed.

The results of assessments for the period of 2011–2013 in Table 2 allow for the following conclusions: energy efficiency indicators at all the facilities analyzed have increased compared with the reference period; on the other hand, we see a variety of trends in dynamics of the integral energy efficiency indicators for some facilities over the period analyzed; E_1 for JSC Ural Copper Plant kept increasing, which reflects negative dynamics at this facility, whereas E_1 for the two other facilities do not show steady dynamics.

5   Conclusion

In general, the proposed methodological tools may be used to solve practical tasks relating to the strategic development of a system and affiliated production units in a context of transition to the "green economy" strategy.

The proposed tools should be used for further research in the following ways: to conduct empirical research for Russian's biggest industrial systems, to work out and

Table 2 Results of energy efficiency assessment for JSC Uralelektromed and UMMC in general

Indicator	2010	2011	2012	2013
JSC Uralelektromed				
Group 1.1. Efficiency of energy-consuming systems and energy saving				
Energy efficiency of gross revenue for primary energy consumption, gram of reference fuel/rouble	23.33 / 1.000	19.90 / 0.853	15.18 / 0.651	15.25 / 0.654
Electricity intensity of gross revenue for primary consumption, kWh/thousand roubles	26.62 / 1.000	24.45 / 0.918	19.16 / 0.720	19.80 / 0.744
Total for group	1.000	0.885	0.684	0.697
Group 1.2. Economic efficiency of energy consumption and efficiency of fixed asset use				
Share of fuel and energy costs in gross revenue	0.09 / 1.000	0.09 / 0.986	0.08 / 0.799	0.08 / 0.841
Coefficient of advance growth of electricity consumption in relation to increase in gross revenue	–	0.92 / 1.000	0.78 / 0.853	1.03 / 1.125
Energy intensity of fixed assets, gram of reference fuel/rouble	31.92 / 1.000	23.15 / 0.725	20.78 / 0.651	19.70 / 0.617
Electricity intensity of fixed assets, kWh/thousand roubles	36.43 / 1.000	28.43 / 0.781	26.24 / 0.720	25.57 / 0.702
Total for group	1.000	0.864	0.752	0.800
Integral indicator at level E_1	1.000	0.875	0.717	0.747
Total for Ural Mining and Metallurgical Company				
Integral indicator E_{total}	1.000	0.845	0.815	0.849

substantiate a system of reference values of energy efficiency indicators for Russia's biggest industrial systems, and to build models of influence of various factors on integral indicators of energy efficiency of industrial systems.

References

1. Krivorotov, V.V., Kalina, A.V., Tretyakov, V.D., Erypalov, S.E.: Methodical tools and results of evaluation of competitiveness of Russian manufacturing complexes (Chapter 3). In: Tatarkin, A.I., Krivorotov, V.V. (eds.) Competitiveness of Socio-economic Systems: Challenges of New Time, pp. 75–138. Economika, Moscow (2014)
2. Ang, B.W., Lee, S.Y.: Decomposition of industrial energy consumption: Some methodological and application issues. Energy Econ. **16**(2), 83–92 (1994)
3. Ang, B.W., Choi, K.H.: Decomposition of aggregate energy and gas emission intensities for industry. Energy J. **18**(3), 59–73 (1997)
4. Energy Efficiency: A Recipe for Success. World Energy Council. http://worldenergy.org
5. Energy Efficiency Indicators: A study of energy efficiency indicators for industry in APEC Economies. Asia Pacific Energy Research Centre. http://aperc.ieej.or.jp/file/2010/9/26/Energy_Efficiency_Indicators_for_Industry_2000.pdf
6. Energy Indicators System: Index Construction Methodology. Washington, DC: Office of Energy Efficiency and Renewable Energy. http://www1.eere.energy.gov/ba/pba/intensityindicators/
7. Energy Policies of IEA Countries. Canada 2009 Review. http://www.iea.org/publications/freepublications/publication/canada2009.pdf

8. Energy Statistics Manual. https://www.iea.org/publications/freepublications/publication/statis tics_manual.pdf
9. Indicators to measure the contribution of energy efficiency and renewables to the Lisbon targets. Fraunhofer Institute for Systems and Innovation. http://www.isi.fraunhofer.de/isi-en/profile/ publikationen.php
10. McKenna, R.: Industrial Energy Efficiency. University of Bath. http://opus.bath.ac.uk/18066/1/ Industrial_Energy_Efficiency_McKenna_030809.pdf
11. Nanduri, M., Nyboer, J., Jaccard, M.: Aggregating physical intensity indicators: Results of applying the composite indicator approach to the Canadian industrial sector. Energy Policy. **30**(2), 151–163 (2002)
12. Worrell, E., Neelis, M., Price, L., Galitsky, C., Zhou, N.: World Best Practice Energy Intensity Values for Selected Industrial Sectors. Lawrence Berkeley National Laboratory, Berkeley. http://eetd.lbl.gov/sites/all/files/industrial_best_practice_en.pdf
13. Anufriyev, V.P., Anufrieva, E.A., Petrunko, L.A.: Increasing competitiveness of regions and companies via green economy. Bull. Ural Fed. Univ. Ser. Econ. Manag. **3**, 134–145 (2014)
14. The OECD green growth measurement framework and indicators. http://www.keepeek.com/ Digital-Asset-Management/oecd/environment/green-growth-indicators-2013_ 9789264202030-en#page21
15. ODYSSEE database. http://www.indicators.odyssee-mure.eu/online-indicators.html
16. CO_2 emissions from the fuel combustion. Highlights. http://www.iea.org/publications/ freepublications/publication/co2emissionsfromfuelcombustionhighlights2013.pdf
17. Tracking clean energy progress 2013. http://www.iea.org/publications/tcep_web.pdf
18. Solomon Benchmarking Greenhouse Gas Services. http://solomononline.com/benchmarking/ further-analysis/greenhouse-gas-services
19. Saaty, T., Kearns, K.: Analytic Planning: The Organization of Systems. Radio and Svyaz, Moscow (1991)

Active Consumers in the Russian Electric Power Industry: Barriers and Opportunities

I. O. Volkova, E. A. Salnikova, and L. M. Gitelman

1 Introduction

Change in the technological basis of the electric power industry by the development of Smart Grid System leads to decentralization of economic decision-making, significant changes in industry management and rules of interaction between economic agents in the market especially households and companies which until recently were only consumers of services. The new technological basis creates conditions for a fundamental change in economic behavior of relevant market agents from "passive" to "active" accompanied by a change in functions and roles of consumer agents and from an agent accepting terms dictated by the electric power system to an "active" consumer who orders services. New roles manifest in demand management actions and the provision of additional system services for load management which gives the consumer an ability to compete with generation.

2 The Concept of Active Consumer in the Energy Sector

The main trend in the development of the Smart Grid is related to the fact that the level of network control and automation increases, technologies become available to a consumer; the consumer becomes not just a subject that consumes electric power

I. O. Volkova
National Research University Higher School of Economics, Moscow, Russia

E. A. Salnikova
Nonprofit Partnership "Market Council", Moscow, Russia

L. M. Gitelman (✉)
Ural Federal University, Yekaterinburg, Russia

© Springer International Publishing AG, part of Springer Nature 2018
S. Syngellakis, C. Brebbia (eds.), *Challenges and Solutions in the Russian Energy Sector*, Innovation and Discovery in Russian Science and Engineering,
https://doi.org/10.1007/978-3-319-75702-5_4

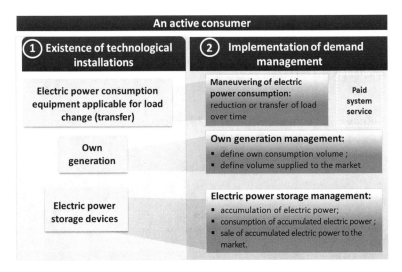

Fig. 1 An active consumer in the energy sector

but begins to play an important role in the energy system by ordering a set of services that he needs. From a subject subordinated by strict regulations and requirements, he becomes a customer. The analysis of international studies and developments on this problem allows us to formulate two main characteristics of an active consumer (Fig. 1): first, technological capabilities and devices that can either generate electric power or use an electric power storage system and second, opportunity and implementation of consumer demand management. The key changes include new consumption patterns. Centralized energy system was focused on large consumers: large steel mills and factories that were not interested in implementing energy efficiency programs and energy saving; now the world trend is the increasing share of consumption of domestic sector, new customers, such as data processing centers (e.g., Google consumes more energy than some industrial consumers).

Thus an "active" consumer is a subject of the electric power market which has a technological capability to change its consumption mode and willingness to participate in demand management programs forming the main characteristics of generated electricity: volume, quality, consumer characteristics, and energy services.

Formulating the above characteristics of "active" consumers determines their main functions in the energy system:

- Coordinated and automatic management of power consumption devices operation mode in accordance with their needs determined by the production plans or household's needs
- Control operation mode of generating equipment and electricity storage systems
- Development of a strategy for participation in the provision of ancillary services [1]

Table 1 Analysis of possibility of realization of "active" consumer functions in the Russian electric power industry (example for household customers)

Scope	Parameter	Now	In perspective
Household consumers	Electric power consumption management	Not implemented or carried out in small volumes in manual mode	Automatic control of the operation mode of electrical appliances (electric isolated devices, for which there is the possibility of changing work schedule: washing machine, dishwasher, air conditioning, heaters, refrigerators, electric device) on the basis of minimizing the cost of energy and satisfaction with the work schedule
	Own generation	None	Presence of own generation (increase in the availability of renewable energy); possibility of transfer its own generation of electricity to the grid
	Electric power storage	None	Availability of electric power storage device
	Strategy	Electrical device mode planning is not carried out or done manually	• Planning of operation mode of each appliance with automatic on/off option for those devices which has a possibility of transferring load in time • Define strategy of own generation loading; consumption and the volume of electric power output to the network • Define strategy of using electric energy storage device: charging, selling of accumulated electricity to the grid, own consumption of accumulated electricity

Our study allows estimating opportunities of implementing "active" consumer functions and behavior strategy of consumers from various fields and economic activities selected based on various power consumption devices' characteristics (Tables 1 and 2).

The analysis shows that in the current environment, the ability to implement the functions of the "active" user is limited due to low availability of electricity storage technologies and small generation and electricity market model specialties: the absence of consumer services market sector and the inability to sell electricity from its own generating capacity to the grid (for small generation). In addition, technological features of some areas of activity result in the inability to implement functions of an "active" user now and in the near future.

Key requirements for an energy system that determine the significance of a consumer "activation" process are the following [2, 3]:

Table 2 Analysis of the possibility of realization of "active" consumer functions in Russian electric power industry (example for industrial customers)

Scope	Parameter	Now	In perspective
Industrial consumers	Electric power consumption management	Reduction in load at peak hours is not considered by the market as an alternative to downloading a backup/peak generation and is not paid by the market	Participation in programs of demand management: • Automatic load reduction in case of emergency • Automatic control of equipment operation modes (load time transfer) on the basis of cost minimization • Provision of system services of reduction in energy consumption to produce the equivalent of the payment in the amount of peak load/backup generation.
	Own generation	No possibility of selling electricity on the market	Possibility of selling electric power from own generation to the network
	Electric power storage	None	Presence of electric power storage devices with large volumes
	Strategy	In some cases (e.g., OJSC "Surgutneftegas") planning of consumption mode and load of own generation is carried out on the basis of market prices	• Planning of equipment operation mode with automatic shutdown option in case of participation in relevant programs with interruption, transferring the load to provide ancillary services to the corresponding payment • Define strategy of own generation loading; consumption and the volume of electric power output to the network • Define strategy of using electric energy storage device: charging, selling of accumulated electricity to the grid, own consumption of accumulated electricity

Motivation of Active Consumer Behavior

This includes creating conditions for independent determination of volume and functional properties (safety and quality) of consumed electric power by consumers in accordance with their needs and grid capabilities based on information on prices, volume, reliability, and quality of electric power.

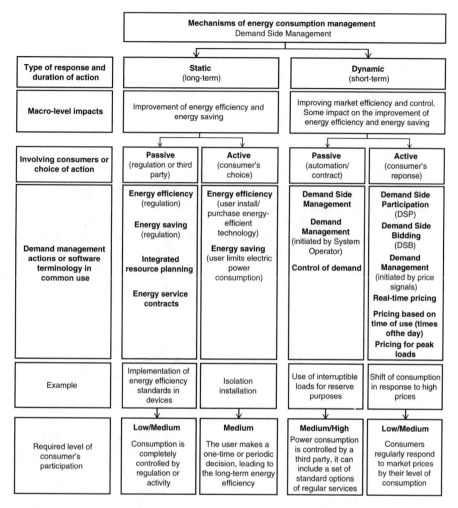

Mechanisms of energy consumption management Demand Side Management			
Type of response and duration of action	**Static** (long-term)		**Dynamic** (short-term)	
Macro-level impacts	Improvement of energy efficiency and energy saving		Improving market efficiency and control. Some impact on the improvement of energy efficiency and energy saving	
Involving consumers or choice of action	**Passive** (regulation or third party)	**Active** (consumer's choice)	**Passive** (automation/ contract)	**Active** (consumer's reponse)
Demand management actions or software terminology in common use	**Energy efficiency** (regulation) **Energy saving** (regulation) **Integrated resource planning** **Energy service contracts**	**Energy efficiency** (user install/ purchase energy-efficient technology) **Energy saving** (user limits electric power consumption)	**Demand Side Management** **Demand Management** (initiated by System Operator) **Control of demand**	**Demand Side Participation** (DSP) **Demand Side Bidding** (DSB) **Demand Management** (initiated by price signals) **Real-time pricing** **Pricing based on time of use (times ofthe day)** **Pricing for peak loads**
Example	Implementation of energy efficiency standards in devices	Isolation installation	Use of interruptible loads for reserve purposes	Shift of consumption in response to high prices
Required level of consumer's participation	**Low/Medium** Consumption is completely controlled by regulation or activity	**Medium** The user makes a one-time or periodic decision, leading to the long-term energy efficiency	**Medium/High** Power consumption is controlled by a third party, it can include a set of standard options of regular services	**Low/Medium** Consumers regularly respond to market prices by their level of consumption

Fig. 2 Mechanisms of demand-side management

Integration of Consumers' Own Generation of Electricity

This includes improvement in the procedure of technological connection and certi-fication of small and distributed generation and power energy storage systems in order to ensure their integration into the power system.

Provide Access of Active Consumers to Electric Power Markets

The ability to implement functions of the "active" consumers in electric power system is provided through mechanisms of power management (demand-side man-agement), which involve a variety of forms of interaction between consumers and other members of the power system. Figure 2 shows the classification of these parameters, developed on the basis of generalization of different terminology and definitions [4, 5].

The first reason for classification of demand-side management mechanisms is the duration of exposure to consumer behavior:

- Long-term perspective: mechanisms to improve energy efficiency
- Short-term: demand management mechanisms (demand response and load management) [6, 7]

Furthermore, one of the important parameters of classification is a type of response:

- Static response (increase in energy efficiency, including through the use of standards) – actions to be taken at any time, depending on the specific signals received from market, or technological system operator requests. Basically, these actions are long-term, for example, the replacement of old equipment with energy efficiency will reduce energy consumption throughout their lifetime.
- Dynamic response – actions to be taken in response to signals received from the market or to predetermined system conditions. Such actions are carried out in accordance with short-term requirements and have a short-term impact (only at run time) although the accumulated effect of the actions of a few consumers can contribute to changing consumer behavior in terms of consumption and market development.

The third parameter of classification is the level of participation of consumer: active or passive. An example of a static response with passive participation of consumers is regulation through establishment of energy efficiency standards of manufactured equipment by the state. This demand management mechanism is the most common nowadays. Active participation of consumers is provided by making a decision, for example, of installation of energy-efficient equipment.

In the case of dynamic response, passive consumer response is influenced by the other participants and is not due to the action of a consumer (e.g., technological operator or energy supply companies, etc.). Most often, these actions are planned and fixed in the contract and are required to maintain the reliability, balancing the energy system or in the case of emergencies. The active participation of consumers implies implementation of proactive actions by the consumer, for example, the transfer of electric power consumption to another time as a reaction to high prices in power system during peak periods. In this case consumers make a choice based on information received from the market and decide to change their behavior in response to said signal. Examples of mechanisms ensuring the implementation of such actions in the electric power market are dynamic pricing and trading on power management. Barriers for implementation of such actions are the concept of management of electric power, focused on producers, and lack of information necessary for decision-making for participants from the demand side [5].

On the basis of these characteristics of an "active" user as well as the analysis of current rules, principles, and technological parameters of an energy system, the authors formulated a system of requirements for the development of energy required for the emergence and integration of an "active" user, which are presented in Table 3.

Table 3 System of barriers to overcome for the emergence and integration of "active" consumers to the power grid

Barriers	Measures (examples)
Technological	
Underdevelopment of electricity accounting and measurement systems as well as information and communication technologies of transferring and processing data	• Development and equipping of power consumer devices with automation systems of remote control modes • Development and implementation of intelligent systems for accounting and measurement allow to monitor the price of electricity in real time • Implementation of information and communication opportunities for two-way interaction between a consumer and energy system using smart metering
Technological complexity of integration of distributed generation to the grid	• Improvement of technologies, methods, and standardization requirements for the integration of the small distributed generation (including renewable energy) to the grid while maintaining the stability and reliability of its work
Economic	
Creating incentives for consumers' "activation"	• Development of motivational management mechanisms • Formation of demand management program system • Dynamic pricing methods • Direct load control methods • Creating a market of system services for a consumer
Organizational	
Need for coordination of consumer retail market	• Agency contracts in terms of selling the potential volume of reduced loading on retail market • Interaction with consumers and aggregation of consumers' offers in terms of costs and volume of potential loading reduction

In this study the authors systematized and highlighted a wide range of components of an effect due to emergence of active consumers. The study shows that this effect is distributed among all market participants. Its appearance in any aspect will be an interesting and positive impact on the development of all sectors of the electric power market (Fig. 3).

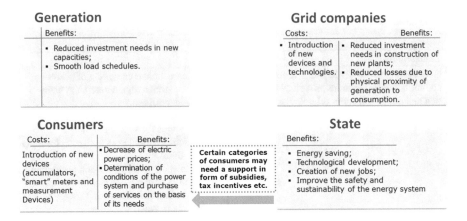

Fig. 3 Effect of "active" consumers' appearance and its components

3 Conclusion

The current situation in the electric power market is characterized by a decreasing efficiency of the industry (lack of competition, extensive practice of "manual" control, etc.) and increasing discontent of consumers on the one hand, and the development of technology on the other creates preconditions for changing the role of consumers in the electric power system from "passive" to "active." This transition calls for revision of management approaches in industry: reorientation from supply-side management, the concept of focusing on cost management of producer companies, to a demand-side management, a concept based on the direct involvement of the consumer in the value creation process.

Switch to a new paradigm of innovation development of electric power industry including integration of "active" consumers involves the following stages:

- Adoption of a strategic decision to move to the industry development based on the intelligent electric power concept fixing that provision in all documents that define the long-term development of the electric power industry
- Development of key provisions of the concept of Smart Grid taking into account the requirements of all participants, technologies for the intelligent infrastructure formation, and legal framework of Smart Grid system, as well as a pilot application of "breakthrough" technologies
- Launch of a new customer-centric electric power market model including the mechanisms of the "activation" of consumers (demand response)

References

1. Volkova, I.O., Salnikova, E.A., Shuvalova, D.G.: Active consumer in an intelligent power industry [in Russian]. Energy Acad. **2**(40), 50–57 (2011)
2. Kobec, B.B., Volkova, I.O.: Innovative Development of the Electric Power Industry Based on the Concept of SMART GRID [in Russian]. IATs Energiya, Moscow (2010)
3. Kobec, B.B., Volkova, I.O.: Smart Grid: Conceptual statements [in Russian]. Energy Mark. **3**(75), 66–72 (2010)
4. Davito, B., Tai, H., Uhlaner, R.: The Smart Grid and the Promise of Demand-Side Management. McKinsey & Company (2010) http://www.calmac.com/documents/MoSG_DSM_VF.pdf
5. Grubb, M., Jamasb, T., Pollitt, M.G.: Delivering a Low Carbon Electricity System. Technologies, Economics and Policy, p. 536. Cambridge University Press, Cambridge (2008)
6. Zgurovets, O.V., Kostenko, G.P.: Effective methods for managing the consumption of electric energy [in Russian]. Online. http://dspace.nbuv.gov.ua/bitstream/handle/123456789/3094/2007_16_St_11.pdf?sequence=1 (2007)
7. Oboskalov, V.P., Panikovskaya, T.J.: Energy consumption management in a competitive electricity market [in Russian]. Online. http://www.sei.irk.ru/symp2010/papers/RUS/S4-14r.pdf

Perspectives on Distributed Generation in the Electric Power Industry

A. Kosygina, I. O. Volkova, and L. M. Gitelman

1 Introduction

Over the last 20–30 years, the electric power sector has tended towards policies favouring liberalization, diversification, decentralization, modernization, and integration. Since the beginning of the 1990s, many countries have introduced reforms into their electric power industries with respect to liberalization, privatization, integrating regional power markets, and attraction of private capital into competitive sectors (generation, etc.).

The development of distributed generation (DG) relies on a wide range of factors, and this very tendency creates new challenges and tasks for power supply systems in the fields of active demand management and implementation of "smart" equipment, advanced energy generation, storage technologies, etc. Development of renewable energy sources (RES), alternative generation technologies, smart energy systems, and distributed generation has increased since the year 2000; furthermore, since 2006, this trend had become one of the more noticeable tendencies in many countries [1–3].

Developing DG in the countries with irregular distribution of population and well-developed heat and electric power systems has some special problems. We consider the Russian power sector as an example. We analyse how DG is being developed in the Russian economy in comparison with worldwide tendencies, its special characteristics, challenges, and obstacles.

A. Kosygina · I. O. Volkova
National Research University Higher School of Economics, Moscow, Russia

L. M. Gitelman (✉)
Ural Federal University, Yekaterinburg, Russia

© Springer International Publishing AG, part of Springer Nature 2018
S. Syngellakis, C. Brebbia (eds.), *Challenges and Solutions in the Russian Energy Sector*, Innovation and Discovery in Russian Science and Engineering,
https://doi.org/10.1007/978-3-319-75702-5_5

2   Main Tendencies for the Development of Distributed Generation in the World

The International Energy Agency (IEA) identifies the following factors in the development of DG in the last decades [4]: development of DG technologies, deferral for upgrades of transmission systems, energy security (diversification of primary energy resources balance and growth in consumer demand for more reliable energy supply channels), environmental protection, and liberalization of the electric power markets.

The leaders in the development of DG are countries where energy policy stimulates this sector, including such directions as renewable energy source (RES), cogeneration, smart grid system implementation, and energy balance diversification (the USA, China, Denmark, Germany, Spain, Japan, etc.). Development of RES and the smart grid system is connected with technological modernization and are not merely the next step for the industry or solely a result of liberalization.

According to the estimation of Global Data, DG capacities will double over the next few years (from 190 GW in 2013 to 389 GW by 2019) [5]. RES generation has priority, specifically solar photovoltaic (PV) generation and wind generation. Also it concerns high-efficiency combined heat and power (CHP) generation. It should be noted that progress in DG development relies heavily on regulator activity, since usually the participants in the "large" electric power system (utilities, companies in generation, transmission sectors, system operator, etc.) do not function as the main drivers in promoting decentralization in the industry. Appearance of new (independent) power producers increases competition in the market and forces down a market share of incumbents, which the older players naturally resist.

The experience of the USA and other countries has shown that there are the following several types of barriers for DG development [1, 4]:

- *Technical*: different technical and juridical requirements (from system operator (SO), transmission/distribution operators (TO/DO)) in order to obtain technical connection to grid or approval for capacity installation. These requirements are too often excessive and sometimes based on obsolete technological decisions.
- *Business practice*: overstatement of price (for capacity reservation for DG or technical connection to grid) and emergence of obstacles for a deal and contract.
- *Regulation*: prohibition by local authorities of DG plant installation (on the grounds of environmental protection or supporting of the centralized electric power system), levying of inconvenient tariffs on RG, etc.

3   Distributed Generation in Russia

The Russian electricity power sector has been in the process of transformation and modernisation for the last 25 years. However this process has been coming to fruition more slowly than in other countries.

The special aspect of the Russian electricity power industry is the geography factor. The country has large and long-distance territory where economy and population are distributed irregularly. About two thirds of the country's territory is outside the centralized electricity power service mainly due to long distances and low density of population, and consequently in those regions, the electric power industry has generally developed in a decentralized manner, relying on small capacity generators generally.

Consequently, DG is present in the Russian electric power sector, though its share was small up until the last decade, when it began to grow, and one of the main trends is the development of consumer generation. The main drivers for this process are problems with reliability of electricity supply and limitation of access to the electric power infrastructure.

It is important to note that development of DG started in the first half of 2000s, and it was before the onset of high price growth in electricity and the completion of the reform of the electric power industry. In the period between 2002 and 2007, the capacity of small and distributed generation (under 50 GW) increased by 25% (2.3 GW) and import increased 12-fold [6]. According to expert estimation in the period from 2012 to 2014, the annual import of equipment for DG (under 25 MW) in Russia was 1.5–2.5 GW. Total expenditure for the import of equipment for DG is more than USD 3 billion [7].

Specificity of Russia DG is developing by consumers of cogeneration on fossil fuels with a capacity of 0.1–25 MW. It can be explained by the cold climate, inappropriate conditions of centralized heat systems in many towns, and the availability of gas supply in the European part of the country. Also, a number of industrial plants produce waste products, which can be used as fuel for electricity or heat generation.

There are no official statistics about DG by consumers. According to expert estimation, DG by consumers grows more rapidly in the industrially developed regions of the South Ural, where industrial plants actively install DG capacities [8]. Additionally, in any cases where consumers switched to "islanding" (isolated operation).

DG is in demand by a wide range of economic agents in the Russian economy from a small to large volume of electricity consumption. First of all, DG is needed in heating systems, in energy-intensive industries (chemical, petrochemical, pulp and paper, iron and steel mills), in transport, and in telecommunications and also by consumers in the isolated and remote territories off the central supply grid.

The advantage of implementing DG in Russia as well as worldwide is its short construction period of 1–2 years, in contrast with the 3–6 years or more required for larger capacity plants. The costs involved in DG projects vary greatly based on a range of factors. Experts' estimates in 2012, evaluated the electricity power price threshold for turning a profit on DG installation projects at about 2.5–3.0 roubles/kW. If the electricity prices, as set by electric power industry, as high, then it is profitable to install own DG plant and expected payback in less than 10 years' time. If consumers install cogeneration capacity and equally utilize both electricity and heat produced, then it is profitable to do so under any market conditions, and the payback period will be 2–5 years.

Therefore, there is no basis to the claim that DG development is just a consumer response to unsuccessful reforms in the electric power industry. The development of DG is a common worldwide tendency, and it started in the Russian economy (as in the USA, EC, other countries and regions) when (1) the electricity power system demonstrated its unreliability or technical limitations and (2) it became possible to realize a commercial project in RG (including projects by consumers). Additionally fast electricity price growth and numerous problems with grid connection have forced consumers to develop DG in the last few years.

Thus, the main causes for the development of DG in Russia are as follows:

- Progress in science and technology and appearance of effective generation (and cogeneration) technologies for small-scale capacities
- Demand for reliability enhancement on local levels (installation of reserve, emergency, and prime capacities)
- High electricity prices for industrial consumers (the main reason for consumer price growth being the rapid increase in tariffs on transmission, rather than wholesale market price growth)
- Obligation for consumers to take part in the capacity market and to be charged additionally (include fees for "must-run" generation capacity)
- Optimization of consumers' energy expenses (electricity and heat)
- Adoption of local energy resources (including RES and waste products)
- Difficulty and duration of procedures for grid connection
- Provision of energy supply in isolated and remote territories
- Uncertainty in the mechanisms of electric power industry regulation and lack of clear development strategy for the current model of the electric power industry (both reasons issue risks for consumers and private investors)

A value of DG (under 25 MW) in the Russian economy can be estimated as 7–15 GW, and its share in total electricity production is 5–8% approximately. Current indirect expert estimates vary greatly from one another, and meanwhile a significant part of micro generation (especially that which produces under 500 kW) remains excluded from official statistics. The potential for development of DG in Russia is substantial, but estimates are quite approximate. According to the evaluation of the Agency for Forecasting of Electric Energy Balance (AFEB), it takes 50 GW of small and medium (under 100 MW) cogeneration capacities for heat system modernization. The potential for RG development in the Russian industry sector is at least 15 GW [9]. Furthermore, there are significant possibilities for DG development in other sectors of the economy (transport, householders, consumer services, etc.). Some experts evaluate the total potential for DG development in the Russian economy to reach 65–90 GW by 2030 [10].

The development of RES generation has become a worldwide trend in the last 7–8 years, but this trend has not been significantly detected in the Russian economy. Neither regulators, nor companies in the generation sector of electric power industry, have expressed significant interest in renewable (and new alternative) energy due to extensive deposits of fossil resources available in the country. Participants in the Russian UES have no incentives for development of new type of prime energy resources, and their reluctance comes across in documents outlining strategies for the industry itself as well as for the fuel and energy sector of the economy as a whole.

Accordingly, these strategies' development of RES in the Russian economy will necessarily occur at a slow rate. Despite this regulators do implement some incentives for RES development in the market zone of unified energy system (UES) such as solar PV and wind plants and small hydrogenation power plants (guarantee of costs recovery for the auction winners by capacity payment mechanism), but participants in the wholesale electricity market, including both producers and consumers, oppose this initiative, since new renewable generation capacities would not only compete with options provided by other electricity power producers but also lay an additional financial burden on consumers.

This conflict of interest is further exacerbated by the current condition of a capacity reserve in the electricity power system surplus. Also it should be taken into account that the GDP growth has dropped dramatically in the last few years. As the situation stands, consumer development of RES projects is mainly limited to enthusiasts operating without any support from regulators. Nevertheless, there remains a stable consumer interest in the development of DG in the following spheres of activity:

- In isolated and remote territories due to high costs of electricity production. A main part of generation in isolated and long-distance territories is thermal power plants, and they use fossil (diesel or coal) fuel. Cost of electricity production is about 15–100 roubles/kWh. In these regions it could be profitable to install hybrid power plants (solar PV and diesel generator, solar PV and wind generator, geothermal plants, etc.).
- In the regions where locally there is fuel RES (biogas, wood waste, etc.).

According to non-profit partnership "Market Council" [11] estimates, the potential for RES generation development in the Russian electricity power industry because of government support and incentives exceeds 25 GW. At the same time according to experts' evaluation, Russia has the resources to develop all types of RES generation and is hypothetically able to meet all its electrical supply in full on the basis of domestic RES resources alone.

In many cases (but not all), development of cogeneration and usage of local prime energy resources for decentralized energy supply by means of DG (including both electric power and heat production) are more effective than installation of large-scale centralized systems (including a generation sector and long-distance transmission networks).

Furthermore, development of DG is beneficial for the country's energy security due to resultant diversification of the prime energy resources balance, and it is beneficial for improving of energy security of a country. It is well known, and many experts have written about these topics, but RG development has been out of strategy documents for industry. It should be noted that according to the Russian Energy Strategy 2030 the share of RG may increase by 15% of total production of electricity and heat.

Spontaneous development of individual generation capacities by consumers introduces additional uncertainty and extends the recovery period for investment projects in the Unified Energy System (UES) and local heat systems. This, in turn, can lead to negative consequences and additional price growth for the consumers who remain tied to these centralized energy systems.

4 Barriers to the Development of Distributed Generation in Russia

Participants of the electric power industry in Russia refer to development of RG mixed (as it is in some other countries), and they are not interested as a whole ring. The industry regulators and NP Market Council do implement measures (with some exception from the rule) to retain large consumers on the wholesale market. If consumers own generation capacity over 25 MW, they are obliged to sell all produced electricity on the wholesale market.

However, as previously discussed, there are barriers and excessive requirements of regulators for grid connection. Their following barriers have risen in the way of DG development in Russia, largely in connection with cooperation with the UES:

- Uncertain rules and barriers inhibiting connection to the UES grid. It is necessary to formalize and standardize regulatory procedures.
- Significant loopholes in regulation and legal framework with respect to DG (among other things – about retail electricity trade) and instability of legislation for the electric power industry.
- Development of technical standards in the UES without attention to DG and the absence of incentive for support of DG development (excluding measures for RES in UES).
- Duration of approval and agreement procedures in regulation bodies and insufficient information support for economic agents (the majority of them are not familiar with possibilities of DG).
- Uncertainty of forecasts for electricity (UES) and fuels prices.

It should be noted that there are not only possibilities but also objective difficulties in developing of DG in the Russian economy and integrating it into the UES; these difficulties are primarily technical.

Additionally, the deterioration of macroeconomic conditions will have a somewhat negative effect on the development of DG on midterm perspective: economic recession, depreciation and high volatility of exchange rates, high inflation rates, and growth of strategic uncertainty in the country development are all likely to adversely contribute to the situation.

5 Supporting Measures for the Development of DG in Russia

Though over the course of the last year the government and regulators have implemented certain measures for supporting DG, this activity has tapered off. In 2011–2012, the government established 32 technological platforms (TPs) to stimulate innovation in the development of the Russian economy. Five of these initiatives concern the electric power industry ("smart grid for Russia", "complex security of industry and energy sector", "environmentally clean effective thermal energy", "perspective technologies of renewable energy", "small-scale distributed energy").

Some proposals for supporting DG were developed and recommended by experts [9, 12]:

- Establishing priorities for installation and sale of electric power and heat for new cogeneration capacities on the market
- Setting a regulatory ban for installation or reconstruction of boiler stations without considering alternative options involving DG
- Simplifying the requirements for technical connection to grid
- Removing retraction to take part in the wholesale market for consumers' generation capacities above 25 MW
- Establishing a stable of legal and regulatory framework
- Developing generic solutions and technical standards for gas-consuming plants in cooperation with gas supply companies

These proposals address the complex set of problems that currently beset the electric power and heat industries. However there are not enough provisions for developing mechanisms for DG in both industries. It is also unclear how and when the logistic developments of municipal power and heat supply schemes will be synchronized, even though legislation has already started to address this issue. Overall, the proposed measures are nevertheless traditional, and they address the development of DG on grounds dictated by the UES rather than encompassing the interests of consumer-generated energy production.

6 Conclusion

Development of DG is one of the visible trends in the electric power industry worldwide. Governments and regulators in many countries facilitate the growth of DG by supporting RES, cogeneration, smart grid systems, "green energy", and increasing competition in the electricity power markets.

The specific case of DG development in Russia hinges on the slow, virtually insignificant integration rate of DG systems into the UES. Consumers install DG actively, but regulators do not support them. Moreover, there are numerous barriers and obstacles to consumers' development of DG. All official measures for promoting DG (including RES) focus on the development of it into the UES and favour generation companies rather than consumers. Despite this, on the whole, the potential for developing DG in Russia is significant. But energy companies, which participate in the UES, along with regulators and the government, currently express little interest in promoting DG in the Russian energy sector. Synchronizing the development of the electric power and heat industries' development with the perspective on DG development by consumers (especially in the industry sector of the economy) constitutes a crucial step on the way to developing DG in Russia.

References

1. Chicco, G., Mancarella, P.: Distributed multi-generation: A comprehensive view. Renew. Sust. Energ. Rev. **13**(3), 535–551 (2009)
2. Volkova, I.O., Salnikova, E.A.: Perehod k intellectualnoi energetike v Rissii: nauchnye i institutsyonalnye aspect. Ekonomika i upravlenie. **5**(55), 77–82 (2010)
3. International Energy Agency (IEA): Technology Roadmaps Smart Grids. OECD/IEA, Paris (2011)
4. International Energy Agency (IEA): Distributed Generation in Liberalized Electricity Markets. OECD/IEA, Paris (2002)
5. Solar Photovoltaic Market the Leading Light in Global Distributed Power Industry; GlobalData, 2014. Online. http://energy.globaldata.com/media-center/press-releases/power-and-resources/solar-photovoltaic-market-the-leading-light-in-global-distributed-power-industry-says-globaldata-analyst
6. Filippof, S.V.: Malaya energetika Rossii. Teploenergetika. **8**, 18–44 (2009)
7. Kiselev, V.: Moderniztsia generiruyuschih objectov i effectivnost' — vzaimosvyazy; NP ACE, 2015. Online. http://energotrade.ru/SpringConferenceMembers2014.aspx
8. Inshakov, O.V., Bogachkova, O.V.: Razvitie maloi raspredelennoi energetiki kak sposob povyhenia energoeffectivnosti i obespecheniya koncurentosposobnosty yjnogo macroregiona i Volgogradskoi oblasty. Vestnik AGTU. Ser. Econ. **1**, 69–76 (2014)
9. Vozmojnosti investirovaniya v promushlennuyu raspredelennuyu generatsiyu v Rossii, ÅF-Consult Ltd, 2013. Online. http://www.np-ace.ru/media/presentations_documents/8POLSUF_summary_report_13082013_final_rus_corr.pdf
10. Voropai, N.V., Trufanov, V.V.: Issledovanie variantov razvitiya EES Rossii na perspectivy do 2030 g. Electro. **3**(13), 2–6 (2013)
11. Nonprofit partnership "Market Council". http://www.en.np-sr.ru
12. Novoselova, O.: Malaya raspredelennaya energetika Rossii: vector razvitiya, problem I dostijeniya, 2012. Online. http://www.smartgrid.ru/tochka-zreniya/intervyu/malaya-raspredelennaya-energetika-rossii-vektor-razvitiya-problemy-i/

Part II
Improving the Management Systems for Sustainable Energy

General Concept for Preventing Energy Crises

L. D. Gitelman, M. V. Kozhevnikov, and T. B. Gavrilova

1 Introduction

Power engineering is a basic infrastructure industry and the backbone of national security. Crises can have detrimental effects on the industry and cause disruption to the security of power supply. Crises can hit the electric power industry in the following ways:

- Demand for electricity (capacity) decreases (one should bear in mind that a decrease in electricity consumption for industrial uses usually occurs later than a drop in output because of continuous power consumption).
- The capacity utilisation factor (CUF) of power plants goes down and production costs go up.
- An uneven loading schedule leads to higher per-unit costs for energy companies because of a drop in consumption by businesses.
- Prices for electricity are subjected to more rigorous regulation in order to curb inflation and provide monetary (government) support to consumers.
- Funding is limited and investment projects have to be abandoned.

As a result, all financial and economic performance results of utilities deteriorate, which might result in a withdrawal of investors from the business, a decline in the quality of equipment operation and maintenance and an outflow of the most qualified personnel.

L. D. Gitelman · M. V. Kozhevnikov (✉) · T. B. Gavrilova
Department of Energy and Industrial Management Systems, Ural Federal University,
Yekaterinburg, Russia
e-mail: m.v.kozhevnikov@urfu.ru

© Springer International Publishing AG, part of Springer Nature 2018
S. Syngellakis, C. Brebbia (eds.), *Challenges and Solutions in the Russian Energy
Sector*, Innovation and Discovery in Russian Science and Engineering,
https://doi.org/10.1007/978-3-319-75702-5_6

The specific nature of crisis impacts on the electric power industry, and the specific role it plays in the economy determines anti-crisis solutions and ways of addressing the negative effects.

2 Anti-crisis Policy Toolkit

Anti-crisis measures can be grouped into:

- Emergency (aimed at mitigating an ongoing crisis)
- Strategic (preventive) ones

It makes no sense to implement the latter when the crisis peaks as they are aimed at improving the sustainability of the sector amid external economic turbulence.

The following emergency anti-crisis measures could be recommended:

1. Heat and power plants, hydro-electric power stations and nuclear power plants should withdraw from the spot market. They should operate at set standard prices (or marginal prices) that match their effectiveness to consumers. Apart from benefits for consumers, pre-established prices (as opposed to volatile spot markets) will help stabilise generators' financial performance.
2. Spot markets should switch over from marginal to average prices (covering all declared costs of selected generators).
3. Tax concessions should be granted to energy companies for keeping excess capacity in operational mode and to the ones who invest in new equipment (including in the form of accelerated depreciation of fixed assets).
4. Companies should conduct an urgent review of their investment projects considering the outlook for funding and procurement capability (with priority given to Russian-made equipment).
5. Grid companies with a monopoly status should have their transmission tariffs set by the regulator on a progressive basis (using reference prices) with a correction factor applied that takes into account the age and condition of equipment. Profit margins should not be aligned with interest rates in effect during the acute phase of the crisis (an option here is to apply pre-crisis parameters). It is also advisable to reduce the tax burden on companies, offer direct government support for urgent investment projects and, possibly, set maximum prices for transmission.
6. Power plants and grid companies should set up cost reduction efforts (devoid of investment), but they should be cautious when making decisions about operating and maintenance personnel levels and their remuneration. Maintenance and repair costs should also be reduced using a scheme that would minimise the possibility of equipment failure.

The following measures could be listed as strategic (preventive) ones:

1. Development of capital-unintensive small-scale distributed generation with different types of generating units, including mobile ones (placed in the closest proximity to consumption clusters)

2. Adoption of advanced demand-side management policies by grid companies and vertically integrated companies and the drawing-up of standard load shedding agreements (as a separate task)
3. Introduction of cost-effective "smart grid" elements for transmission and distribution infrastructure and installation of smart metres with the two-way communication capability
4. Reinforcement of grid interconnections (including the use of direct current) to validated levels, considering the possibility of merging grid companies and electricity retailers and working on the merger details that cover cost and price reduction as well as combined responsibility for security of the electricity supply in regions
5. Forward-looking innovative development of the machine building and electrical manufacturing industries in order to fully provide new energy facilities and those under refurbishment with locally made equipment; provision of comprehensive after-sales support (including all kinds of repair) by the manufacturer during the entire service life

In today's crisis context, emergency measures are being taken across the board and widely discussed at all levels, generating general approaches and compromise solutions, while strategic measures have been put on the back burner. They are only occasionally mentioned when the current situation is being discussed. Such measures, however, provide for sustainable development of the industry and make it possible to alleviate crisis in the future. One should not, therefore, forget about strategic solutions and supplementary tools when going through deep crisis. Among such always useful tools are:

- Demand-side management
- Development of small-scale power generation
- Cost management
- Asset management

The first two of them will be further considered below. Cost management and asset management will be elaborated on in other chapters of this book.

3 Demand-Side Management

Evidence on the development of power engineering globally shows that a long-term balance between capacity and load growth can only be secured through rationalisation of electricity consumption that is referred to as "demand-side management (DSM)". DSM programmes could be initiated by consumers, who are challenged by competition to reduce their production costs, as well as by electricity suppliers seeking to cut the growing cost of building new generation and transmission capacity [1].

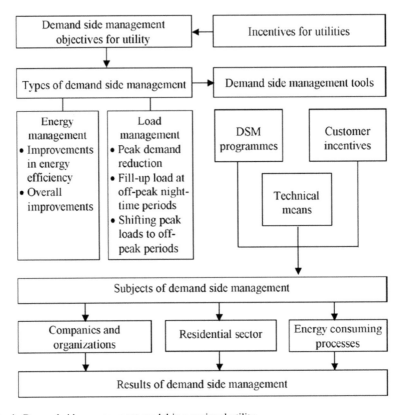

Fig. 1 Demand-side management model in a regional utility

Essentially, demand-side management means that the electricity supplier implements a planned and targeted policy of controlling the quantity, patterns and timing of electricity consumption in its service region. At the same time, the company views a growth in power usage effectiveness and the development of its generating (grid) capacity as complementary ways of ensuring the power supply to its consumers. Pro-active management of demand for electricity and capacity makes it possible to meet emerging energy needs of any local industry at minimal cost, which is particularly important in a time of crisis and during the recovery period.

Apart from the utility as the control agent, demand-side management as a system includes forms, means, tools and control subjects. The effectiveness of demand-side management is determined by the end results that are different for utilities, consumers and the region as a whole (Fig. 1).

The utility benefits by saving on investment and operational costs, expanding its market, increasing the long-term stability of its financial performance and creating a favourable image for itself in the region.

Consumers enjoy increased reliability and quality of the electricity supply as well as lower and stable electricity and heating tariffs and make direct savings by reducing the energy intensity of their products.

From the perspective of long-term societal interests, *the region* benefits from more stable energy supply, both during the crisis and economic recovery, a higher level of energy self-sufficiency and cleaner environment.

A general methodological approach to testing the cost effectiveness of demand-side management programmes is by comparing forecast capacity-related and energy savings (in monetary values) to overall costs of developing and implementing a programme [2]. The monetary value of energy efficiency is established through the prism of avoided costs of building new generation capacity, i.e. as the opportunity cost of electricity production. It is recommended that the most cost-effective installations should be adopted as alternatives for comparison, renewable energy installations and natural gas-fired power plants being the most typical choice. The programme costs should encompass all appropriate capital and operating expenditures, including discounts on prices and tariffs except annual expenditures on servicing earlier investments.

4 Distributed Generation

As a rule, small-scale power plants service consumers who cannot be connected to electrical grids for technical reasons. In the majority of cases, the capacity of such installations varies between several kilowatts (a detached house) and a few dozen megawatts (a village with a population of several thousand, or a small company). However, consumers with access to centralised grid-connected power express an interest in such generators, too. They are attracted by the reliability of stand-alone power systems, cost-related and environmental benefits and the possibility of increasing their energy security by using locally available energy sources [3].

A short and cost-effective investment cycle (including design, construction, installation and setup), low maintenance costs and ease of operation make small-scale power generation a highly attractive area of power engineering (provided there is no discrimination as regards access to markets and guaranteed sales of electricity and heat).

Technological flexibility is a noteworthy feature of distributed generation. Generation units are compact, easy to deliver and install in almost any location and are fully automated, which makes it possible to promptly put them into service in places with a shortage of generation or where access to the grid is limited. Being essentially a type of local electricity generation, small-scale generation options provide an opportunity to overcome technological and economic inertia that big energy systems are known for. This makes distributed generation a universal tool for preventing electricity supply crises in regions.

The owners of small-scale installations come to appreciate their technological flexibility in the course of operation. For example, small generators can be switched on during peak hours and put on standby when prices are low. The ease of installation makes it possible to quickly increase capacity in response to favourable market conditions. Some distributed generation installations are so mobile that they can virtually "follow the market".

The development of small-scale power generation reduces transit flows, takes load off the grids and increases the transfer capacity of transmission lines. Additionally, spending on main and distribution electricity networks and substations go down, as do the costs of refurbishing overloaded transmission lines and grid losses.

The growth in small-scale generation as private business encourages competition, especially in local retail markets. Independent power producers will be a source of growing pressure for regional utilities, encroaching on their cash flow.

At the same time, small-scale power generation is confronted with a number of problems, the prime one being a relatively high cost of some environmentally friendly installations. This calls for scientific and technological advances in small-scale generation.

It has to be pointed out that some installations (e.g. windmills, solar farms and some heat and power plants) are unable to operate in a load dispatch control mode. This imposes restrictions on utilising their reserve capacity and raises the issue of reliability of the electricity system. A possible solution could be to create powerful electric energy storage devices.

Small generation units are installed in distribution networks in close proximity to specific consumers. This changes the technical characteristics of the distribution network and can affect its stability. To avoid such effects, energy systems need to be modified to be able to incorporate such installations.

The government should provide legal, technical and economic conditions for promoting small-scale power generation. It is particularly important to introduce uniform rules for a small-scale power generation market that would:

- Govern access for independent power producers to the grid.
- Introduce regulation procedures for electricity and heat prices.
- Establish power supply arrangements (govern relations between independent producers, their consumers, grid companies and electricity retailers).
- Establish dispatch control procedures.
- Introduce a mechanism for encouraging investment in small-scale power generation.

Economic regulation should primarily aim to support temporarily uncompetitive power plants such as heat and power plants during periods of an unfavourable ratio of natural gas prices to electricity (heat) prices, some renewable energy installations and microgeneration units powered by natural gas and liquid fuel. Incentives should be provided at all cost-intensive stages from construction and grid connection to operation. The mix of possible incentives is well known and includes interest rate subsidies, tax concessions, accelerated depreciation, operating cost subsidies and sales guarantees.

For example, the European Union has been implementing renewable energy support programmes (primarily to reduce CO_2 emissions). The programmes might differ in specific countries, but generally they combine a system of emission allowances with pricing instruments.

5 Conclusion

The implementation of the above-described measures will eventually make it possible to:

- Mitigate the effects of an economic crisis on the electric power sector (thanks to "emergency" measures).
- Avoid the impact of crisis occurrences by taking "organisational and technical" (strategic) measures.
- Find a solution to the "compromise" price issue in extraordinary circumstances.
- Ensure required stability of the electricity supply to business consumers and households (the issue of acceptable prices for electricity is to be addressed by the regulator at government level and at government expense).
- Preserve and boost workforce potential of the industry, retain investors and carry out refurbishment projects even in unfavourable circumstances.
- Ensure national energy security through import substitution for refurbishment and construction purposes.

Acknowledgement The work was supported by Act 211 Government of the Russian Federation, contract № 02.A03.21.0006.

References

1. Gitelman, L.D., Ratnikov, B.E., Kozhevnikov, M.V.: Demand-side management for energy in the region. Econ. Reg. **2**, 71–78 (2013)
2. Understanding Cost-Effectiveness of Energy Efficiency Programs: Best Practices, Technical Methods, and Emerging Issues for Policy-Makers. http://www.epa.gov/cleanenergy/docu ments/suca/cost-effectiveness.pdf
3. Gitelman, L.D., Ratnikov, B.E.: Economics and Business in Power Engineering [in Russian]. Ekonomika Publishing House, Moscow (2013)

Managing Productive Assets of Energy Companies in Periods of Crisis

L. D. Gitelman, M. V. Kozhevnikov, and T. B. Gavrilova

1 Introduction

The main anti-crisis considerations for energy companies' productive assets [1–4] are associated with optimising their non-core assets structures; improving production logistics, extending leasing schemes; or with indirect economic measures of an operational nature (e.g. improvements in stock turnover). Meanwhile, owing to the high level of inertia of the energy industry, it is necessary to develop a system of asset management which could provide, in times of crisis (characterised by limited investment opportunities and the general uncertainty of the external environment), continuing stable operations, including power supply reliability.

For developing countries, the following industry peculiarities should be taken into account in the course of asset management system development:

- The high depreciation of fixed assets, both in the generation and distribution sectors
- The lack of a developed economic mechanism for managing reliability
- The noticeable gaps in implementing technical regulations

L. D. Gitelman · M. V. Kozhevnikov (✉) · T. B. Gavrilova
Department of Energy and Industrial Management Systems, Ural Federal University,
Yekaterinburg, Russia
e-mail: m.v.kozhevnikov@urfu.ru

© Springer International Publishing AG, part of Springer Nature 2018
S. Syngellakis, C. Brebbia (eds.), *Challenges and Solutions in the Russian Energy Sector*, Innovation and Discovery in Russian Science and Engineering,
https://doi.org/10.1007/978-3-319-75702-5_7

2 Principles of Productive Asset Management

Current models of asset management focus mainly on solving the problem of cost optimisation (i.e. reduction) in relation to maintenance and repair [5], as well as managing equipment failure risks. Indeed, some companies are turning to models that are based on certain ratios of reliability and risk, namely, high reliability (at higher costs) with low risk and vice versa.

Such asset management systems should also prevent or minimise the negative effects of such types of risks such as increases in fixed assets' depreciation or the ageing (deficit) of qualified personnel. According to the study carried out by the foreign consulting firm Deloitte [6], a systematic analysis of productive assets' status allows energy companies to reduce the proportion of expensive, "reactive" (force majeure) repairs in favour of preventive maintenance (Fig. 1).

It is appropriate to note that personnel ageing is often considered as a technological (rather than an organisational) risk. This trend is relevant for the energy industry in many countries, for example, according to the study "Development Strategy of the US Energy Industry" made in 2014 by the Black & Veatch group of companies, personnel ageing is among the top ten strategic challenges for the industry.

More progressive energy companies refuse to focus on reliability as the main performance indicator and begin to implement complex strategic schemes in which asset management goals and objectives are determined on the basis of the companies' development strategy. This is believed to help achieve a balance between long-term and short-term goals of the energy companies (in terms of reliability) while meeting the interests and requirements of all participants in the electric power market. This approach is widespread in developed countries and is gaining popularity in developing countries. It is based on various international asset management standards: for example, standard PAS 55 (Fig. 2).

Different IT tools and software complexes are used in asset management. They help carry out remote monitoring of equipment's status; predict the behaviour of the equipment, and its individual components, throughout the asset life cycle;

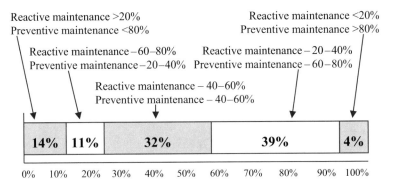

Fig. 1 Reactive and preventive maintenance distribution in energy companies choosing risk management strategies to manage assets

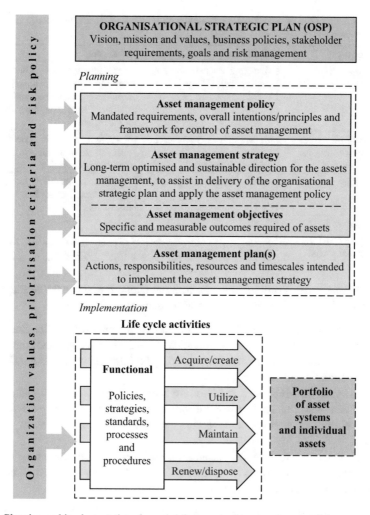

Fig. 2 Planning and implementation elements of an asset management system [7]

automatically collect data on failures; and so on. The present asset management system is therefore considered to be intelligent [8, 9]. The table below provides the characteristics of such intelligent systems (Table 1).

3 Energy Companies' Asset Management Systems

To prevent negative effects in times of crisis, the concept of managing energy companies' productive assets should be based on the following provisions:

Table 1 Comparison of traditional and intelligent systems of managing energy companies' productive assets

Characteristics	Traditional	Intelligent
System essence	Managing assets taking into account their life cycle	Managing assets taking into account their life cycle, the company's long-term strategic interests, and the possibility of immediate reaction to force majeure
Degree of information provision for the asset management process	Moderate, mainly concerning ways of collecting data on equipment operation	Very high, integrating intelligent decision-making processes concerning operations with assets
Asset maintenance system	Mainly planned and predictive maintenance	Maintenance according to equipment technical condition information
Connection with the company strategy	Weak	Strong
Management focus	Operational effectiveness	Risk management

1. The Management Diagram Consists of three Units: Intracompany Management, External Management, and Demand Management.

The intracompany unit is designed to implement the commercial interests of energy companies. These are the companies' growth, increasing their competitiveness in the electric power and heating markets, and fulfilling contractual obligations to their customers and partners.

The external unit is subdivided into two sectors: a system of technical, technological, economic, financial, and environmental norms, guidelines, and standards and the norms, regulations, and operational instructions of supervisory control bodies.

The *demand management* unit provides for cooperation between energy companies and consumers in the area of increasing energy efficiency and the rationalisation of their modes (i.e. load curves).

2. Both fixed and working assets are considered to be the focus points of management.
3. The management processes involve all stages of the productive assets life cycle, that is, formation, usage, and reproduction.
4. The levels of achieving standard and planned indicators for meeting electric power and heating demand (launching new facilities), for the equipment's readiness to bear the load, as well as for production reliability and safety and its energy efficiency are considered to be the ultimate asset management outcomes.
5. The financial reproduction aspect of managing fixed assets is represented by an investment-efficient depreciation mechanism [10].

The system of managing energy assets includes three subsystems (units):

- The information unit (monitoring the productive assets status, information about changes in the rules of energy markets' operation, as well as changes of industry standards, and decisions in the field of technical and fuel policy in the industry)

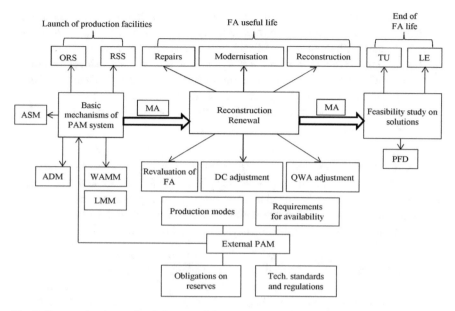

Fig. 3 Structural and operational diagram of the management of the productive assets in an energy company

- An innovation unit (organisational economic and technical solutions developed and implemented at different stages of the productive assets life cycle)
- A regulatory and measuring framework for managing energy assets (a set of indicators that perform analytical (monitoring), design, innovation, evaluation, and stimulation functions)

Below is a diagram illustrating the management of productive assets in an energy company (Fig. 3).

Captions in the figure: ADM, (fixed) assets depreciation mechanism; ASM, assets status monitoring method; DC, depreciation charges; FA, fixed assets; LE, life extension; LMM, logistics management mechanism (fuel supply); MA, assets monitoring; ORS, operational reliability system; PAM, productive asset management; PFD, power facility demolition; QWA, quota for working assets; RSS, repair service system; TU, technical upgrade; WAMM, working asset management mechanism.

Managing the key component of productive assets – fixed assets – is implemented by performing the following functions:

1. *Creation of fixed assets.* This covers the entire investment cycle from the design to launch stage (including the breaking-in period). This function appears to be crucial. First, it is during the investment cycle that the power facility's effectiveness potential (generally understood as reliability, safety, profitability, and durability) is established. This potential is unlocked at the exploitation stage and is maintained and even increased through fixed assets restoration and renewal.

Second, power facilities construction requires customers to work with a large number of organisations; therefore, at every stage of the investment cycle, they have to deal with the problem of selecting an effective contractor.

2. *Assessment and timely elimination of equipment wear and tear.* Depending on its scope and investment restrictions, the accumulated *physical deterioration* is eliminated by means of repairs (major overhauls), by extending the service life beyond normal economic life, and, finally, by means of replacement (renovation) of the equipment.

Functional depreciation (obsolescence) is made up for by upgrading the equipment and by rebuilding and overhauling the enterprise (or the power generating unit). It ought to be noted that economically reasonable repair and fixed assets renewal activities should be aimed at implementing two key management objectives: firstly, to ensure availability and reliability of the equipment' and, secondly, to raise the production process energy efficiency to competitive values.

3. *Classification of fixed assets.* Facilities are combined into separate groups depending on their industrial purpose, technical specifications, and service lives. Although there are common (normative) classifications according to industries, they can be specified within particular energy companies. This classification allows for tracing the fixed assets structure dynamics and for the development of solutions for its improvement resulting in an increase in the returns on these assets.

4. *Defining the value of fixed assets.* There now exist different ways and methods to value the fixed assets of enterprises. Of particular importance is the balance valuation of the current (market) cost, taking into account the accumulated depreciation, that is, at the real "physical-technical" level.

5. *Formation and adjustment of the mechanism for fixed assets depreciation.* Depreciation functions include not only full compensation for physical and "moral" depreciation of fixed assets but also refunds of the investments in depreciated assets to the owners. Moreover, the depreciation fund is the enterprise's own investment source for the renovation and extension of fixed assets (i.e. of the company's value).

6. *Development and application of a methodology for assessing the effectiveness of fixed asset management.* To effectively manage any object, it is necessary to measure and interpret the results of actions performed and solutions implemented. It is therefore necessary to establish a system of performance indicators that characterise all the main aspects of energy companies' asset management. These indicators must have a rational degree of aggregation, be sufficiently informative, and be easily calculated. The next step is to create, on the basis of these evaluation characteristics, an automated data processing system based on IT technologies.

4 Conclusion

The global trend of market mechanisms spreading in the energy industry implies a qualitative transformation of energy companies' productive asset management systems. This study shows that currently the leading energy companies are shifting to intelligent asset management with flexible information systems that provide constant "equipment–personnel" intercommunication, analytical models, and algorithms for decision-making in relation to assets, and so on.

Energy companies' experience shows that introducing asset management systems has significant economic benefits. These benefits include reduced equipment repairs and maintenance costs, reduced volumes of material and productive assets, and increased cost-effectiveness of production (itself especially important in times of crisis due to limited investment resources). Advanced asset management systems also facilitate predicting the probability of equipment failures and reduction of the risk of *force majeure*.

Acknowledgement The work was supported by Act 211 Government of the Russian Federation, contract ☒ 02.A03.21.0006.

References

1. Deberdieva, E.M.: The transformation of productive assets structure in oil and gas companies: Preconditions and factors [in Russian]. Manag. Econ. Syst. Elect. Sci. J. **3**, (2015.) http://www.uecs.ru/uecs-75-752015/item/3386-2015-03-03-08-07-30
2. Petrova, Y.M.: A company working capital management in crisis conditions [in Russian]. J. Econ. Integ. **1**, 103–105 (2009)
3. Gubin, V.A.: On instrumental provision of economic system management methodologies in a crisis environment [in Russian]. Polythem. Netw. Elect. Sci. J. Kuban State Agrarian Univ. (KSAU Sci. J.) **2**(76), (2012.) http://cyberleninka.ru/article/n/ob-instrumentarnom-obespechenii-metodologii-upravleniya-ekonomicheskoy-sistemoy-v-usloviyah-krizisnoy-ugrozy
4. Contemporary approaches to asset investments in the energy industry Assessing the demand, limitations, structures and approaches for optimal energy asset investments. (2012). http://www.accenture.com/SiteCollectionDocuments/PDF/Accenture-Contemporary-Approaches-Asset-Investments-Energy-Industry.pdf
5. Maslov, A., Frolov, K., Volkova, I.: Analysis of world experience in building asset management systems in energy companies [in Russian]. Energy Mark. **7–8**, 31–36 (2007)
6. Asset management: A risk-based approach. http://www2.deloitte.com/cy/en/pages/energy-and-resources/articles/risk-based-approach-benchmark-survey.html
7. PAS 55–1:2008. Part 1: Specification for the optimized management of physical assets. http://www.mop.ir/portal/File/ShowFile.aspx?ID=840ee27f-1f65-4f15-a7df-2f25dfc83d51
8. Asset management in the utilities industry. http://www.ibm.com/expressadvantage/br/downloads/Asset_management_in_the_utilities_industry.pdf
9. Schneider, J., Gaul, A., Neumann, C., Hogräfer, J., Wellbow, W., Schwan, M., Schnettler, A.: Asset management techniques. http://pscc.ee.ethz.ch/uploads/tx_ethpublications/fp1000.pdf
10. Gitelman, L.D., Ratnikov, B.E., Kozhevnikov, M.V., Simonov, M.A.: Basics of Energy Companies' Asset Management: A Practical Guide for Managers [in Russian]. UrFU, Yekaterinburg (2012)

Cost Management Methods in Energy Companies

L. D. Gitelman, M. V. Kozhevnikov, and T. B. Gavrilova

1 Introduction

Cost management is one of the most important components of business management at all stages of its development and under any external conditions. While technical development, modernization, and increasing the value and investment attractiveness of business are a priority during periods of economic growth, during periods of crisis managers tend to pay more attention to costs. In a situation when profit growth is impossible and access to financial resources is restricted, cost reduction becomes the main strategy for a company's survival.

Figure 1 features a general scheme of cost management in the electric power sector. In accordance with strategic goals set for the industry, cost management envisages a wide range of actions aimed at drastically increasing the technical level and building up-to-date infrastructure compliant with the world's standards as well as at cost reduction and boosting energy efficiency in the course of daily operations. In order to cope with the tasks, the industry has a wide range of trialled and tested techniques that require information, technical and organizational support.

L. D. Gitelman · M. V. Kozhevnikov (✉) · T. B. Gavrilova
Department of Energy and Industrial Management Systems, Ural Federal University,
Yekaterinburg, Russia
e-mail: m.v.kozhevnikov@urfu.ru

© Springer International Publishing AG, part of Springer Nature 2018
S. Syngellakis, C. Brebbia (eds.), *Challenges and Solutions in the Russian Energy
Sector*, Innovation and Discovery in Russian Science and Engineering,
https://doi.org/10.1007/978-3-319-75702-5_8

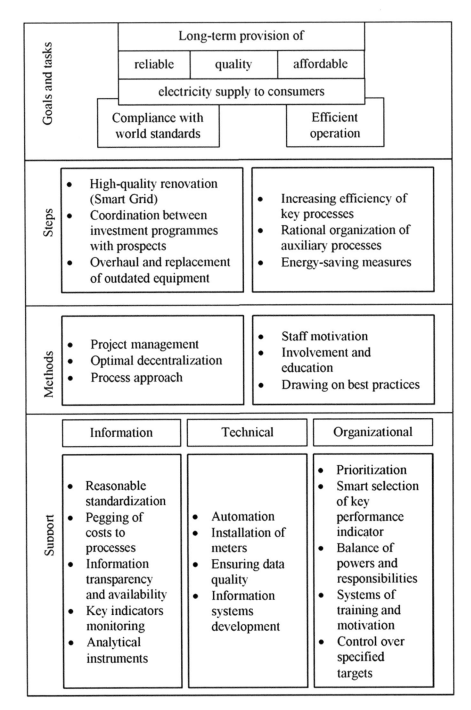

Fig. 1 Cost management in the electric power sector

2   Strategic Solutions for Cost Management

Projects of modernization of the sector based on the Smart Grid concept, that is, the creation of a unified energy information system oriented toward ultimate customer needs and capable of self-regulation and self-healing, have the biggest potential for the improvement of the cost management system [1–5]. Decentralization, intensive two-way information exchange, diagnostics and automated real-time problem solution provide Smart Grid-based energy systems with considerable advantages.

At present, the idea of Smart Grid in Russia is regarded as a concept of an intellectual active and adaptive network characterized by the following [6]:

- The network has a sufficient number of active elements capable of changing its topological parameters.
- The network has a lot of detectors to measure current operating parameters in order to assess the network's condition in various regimes of energy system operation.
- The network has a data acquisition and processing system (hardware and software systems) as well as devices to control the network's active elements and customers' electricity-generating equipment.
- The network has all required elements and mechanisms enabling it to change its topological parameters in real time as well as to interact with adjacent energy facilities.
- The network has tools for automated assessment of the current situation and making forecasts for the network's operation.
- High speed of response of the control system and information exchange.

Operational and maintenance costs in an energy system based on the Smart Grid concept decrease due to the following reasons:

- The incidence of power failures goes down, electricity bills are reduced and staff has to respond to power failures less often, thanks to automated switching off and switching on.
- Scheduled maintenance is replaced by condition-based maintenance, thanks to real-time monitoring of assets.
- The risk of overload is reduced, thanks to the use of up-to-the-minute information on the condition of assets with the help of Smart Grid monitoring technologies.
- Electricity distribution losses decrease by over 30% due to the optimization of power plant performance and balancing of the energy system.

It should be emphasized that the Smart Grid concept focuses on the revision of principles of development and creation of a new innovative technological platform of the electric power sector rather than merely on the modernization of some technologies and equipment.

3 The Use of Up-to-Date Analytical Tools

In a context of crisis, strategic goals remain unchanged, but measures aimed at achieving the goals tend to change significantly. The unavailability of financial resources hampers the implementation of big projects capable of taking the energy sector to an absolutely new level. Modernization, replacement of outdated equipment and efficient organization of repair and maintenance become ever more important. Predictive analytics helps to ensure proper implementation of the above-mentioned tasks.

Predictive analytics enables to reveal risks before the problem becomes reality, to take preventive actions and, consequently, to reduce losses and save resources.

Prediction of equipment failures is one of the most typical applications of predictive analytics in the manufacturing sector. Prediction of failures envisages the use of complex analytical models, creation, verification and actualization of which is achieved with the help of specialized IT solutions. The SAS Institute and IBM are the recognized leaders in this segment (Fig. 2).

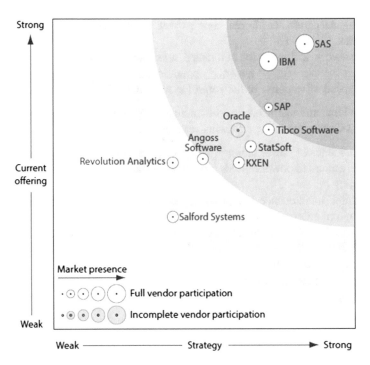

Fig. 2 Positions of IT companies in predictive analytics market [8] (Source: Forrester Research, Inc.)

According to data of the US Department of Energy, the introduction of predictive technical maintenance of equipment results in a significant economic effect:

- Maintenance costs go down by 25–30%.
- The number of failures is reduced by 70–75%.
- Downtime is down 35–45%.
- Output can increase by 20–25%.

Evidence from energy companies proves high efficiency of using predictive analytics for the diagnostics of defects and improving repair and maintenance processes. The roll-out of SAS predictive analytics at a gas processing plant with the annual capacity of 10 million cu. m. of associated gas made it possible to eliminate serious failures in the gas desulfurization system. Standard analytical tools failed to identify the cause of those repeated failures in the system. The oil separator broke down 22 times over 24 months. Unscheduled outages and repair works would result in the interruption of the production cycle. The situation was complicated by the fact that technical maintenance was not always possible immediately after the separator failure. Transition to diagnostic repair with the use of SAS predictive analytics ensured steady uninterrupted operation of the separator for 36 months. The analytical model made it possible to learn about a potential failure of the gas compressor unit 75 days ahead, and preventive repair was conducted. Analytical tools revealed the cause behind repeated failures of the electric centrifugal pumps: temperature fluctuations, albeit within allowed ranges, caused the problem. Equipment data analysis made it possible to extend intervals between repairs and to significantly reduce technical maintenance costs [7]. Having a vast experience in analytical tools roll-out, the SAS Institute offers a wide range of solutions to increase the efficiency of technical maintenance and repair (Fig. 3).

Demand forecasting is another important trend in the energy sector. The use of IBM SPSS PA for demand forecasting by CIPCO, a major energy supplier, enabled it to improve planning of energy purchases and price policy, to conduct non-stop monitoring of efficiency, to monitor the grid condition and to improve its investment programme.

4 Sample Solutions

Prognoz, a Russian company that provides solutions in predictive analytics, has developed and introduced a system of monitoring, analysis and forecast for the electric power sector of one of China's regions. The project was commissioned by a regional branch of State Grid Corporation of China (SGCC) – China's biggest electric grid company, which ranks seventh in the Fortune Global 500 list. The company specializes in construction and use of electric grids in China and abroad. SGCC has a monopoly in electricity transmission and sales.

The solution relies on a system of models for computing future energy consumption volumes based on short-term and long-term forecasts for the region's

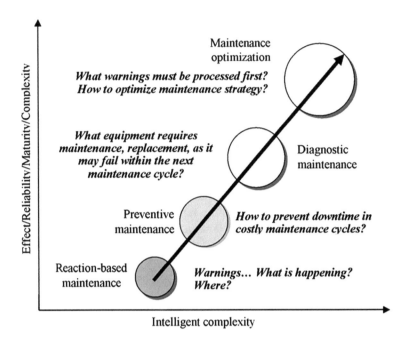

Fig. 3 Possibilities for boosting technical maintenance using analytical tools [9]

development as well as on data on electricity production and consumption. A model of the electric power sector and a model of the regional economy, which describes the economy of a given region with its peculiarities, are components of the system. For example, the regional model takes into account parameters of the state and regional policies (investment growth, foreign currency rates, etc.) as well as global economy indicators. This model calculates the key indicators of the regional economy based on one or several scenarios, which allows for considering various regional development alternatives.

The indicators obtained with the help of this model (for instance, gross value added, personal incomes, capital investment, prices) are used for creating models of the electric power sector which make it possible to forecast electricity consumption in the region and for major groups of consumers. In addition, a forecast is generated for a number of indicators in the electric power sector, such as sales of electricity and capacity utilization.

The project was completed in late 2013. The deployment of the system made it possible to reduce the costs and time for obtaining real-time data as well as to improve information quality and speed up the solution of analytical and projection tasks. In addition, the move increased the accuracy and substantiation of decision-making, thanks to the use of industry development forecasts.

Prognoz also developed and introduced a software package to model conditions in the wholesale market for electricity and capacity for Inter RAO, which enabled the company experts to obtain plausible assessments of demand from their main customers by changing input parameters.

For the benefit of the customer, some 12,000 short-term, mid-term and long-term models were developed for 74 regions and 5 types of economic activities. A consensus forecast module was created. Visualization of data with the help of a map, pie and bar graphs with adjustment options for relevant data on each of them enables the user to conduct quality analysis of both forecast data and the actual one. The user can select relevant indicators from the database and get a graphic representation for certain indicators.

The system offers a graphic representation of the entire regional model of conditions in the Russian wholesale market for electricity with an option of viewing data for each input and resultant indicator. It is also possible to view data for all models of electricity demand forecasts as well as to view models for auxiliary and resultant indicators for each region. Both rapid and routine reports are available, enabling operational analysis of main modelling indicators with an adjustment capability for data presentation.

5 Conclusion

Prompt and flexible reaction to changes in external conditions, adaptability and flexibility of management at all levels, along with active use of forecasting tools become essential amid crisis. Decentralization, removal of internal barriers and exchange of best practices are prerequisites not only for the system stabilization but for survival.

For its part, the organizational component gets an additional impetus due to an increase in personnel's activity. Proper motivation and a well-designed system of training make it possible not only to reduce losses but also to significantly increase employees' involvement in the process of seeking and mastering effective methods of accomplishing tasks. The improvement of staff's skills amid crisis provides a stepping stone to the company's growth and development when the economy stabilizes.

Acknowledgement The work was supported by Act 211 Government of the Russian Federation, contract № 02.A03.21.0006.

References

1. Gitelman, L.D., Ratnikov, B.E.: Energy Business, [in Russian]. Delo Publishing House, Moscow (2008)
2. Kobets, B.B., Volkova, I.O.: Innovation Development of Electric Power Sector Based on SmartGrid Concept [in Russian]. IATs Energiya, Moscow (2011)
3. Vaynzikher, B.F. (ed.): Electric Power Sector of Russia 2030: Target Vision [in Russian]. Alpina Business Books, Moscow (2008)

4. European Technology Platform Smart Grids, Vision and Strategy for Europe's Electricity Networks of the future; European Commission Directorate-General for Research Information; Communication Unit European Communities; European Communities, http://ec.europa.eu/research/energy/pdf/smartgrids_en.pdf (2006)
5. Grids 2030. A national vision for electricity's second 100 years; Office of Electric Transmission and Distribution of USA Department of Energy. http://www.ferc.gov/eventcalendar/files/20050608125055-grid-2030.pdf (2003)
6. Dorofeyev, V.V., Makarov, A.A.: Active adaptive network is Russia's energy system's new quality, [in Russian]. Energy Expert. **4**(15), 28–34 (2009)
7. Gitelman, L.D., Ratnikov, B.E., Kozhevnikov, M.V.: Demand side management for energy in the region. Economy Region. **2**, 71–78 (2013)
8. Gualtieri, M., Curran R.: The Forrester Wave™: Big Data Predictive Analytics Solutions, 2015. Online https://www.victa.nl/alteryx/wp-content/uploads/The%20Forrester%20Wave%20Big%20Data%20.pdf
9. Mitroshkina, V.: Application of SAS for technical maintenance optimization [in Russian]. Online http://www.it.ru/activity/activities_archive/4021/

Improvement of the Management System in a Power Grid Company with ERP Systems Implementation

A. Y. Makarov, O. Y. Antonova, R. I. Khabibullin, and A. S. Semerikov

1 Introduction

Improvement of the management system primarily demands accurately formulated requirements to the forthcoming changes. The following goals of the improvement program of the management system were set in a company.

1.1 Unitized and Standardized Work Performance

The structure of the company includes ten power grid branches usually performing an identical set of functions. Regulatory documents (RD) were formed on the basis of unification and standardization of the results of the work of production departments; thus, regulatory documents were able to distribute and to demand its fulfillment.

A consolidated plan of RD is approved on a yearly basis to regulate production and administrative activities (Table 1).

Uniform procedures, principles and rules of building and supporting of a regulatory documents management system aimed at continuous and appropriate performance of all business processes, are established. The result of the carried-out work includes the planning procedure of creation, development, coordination, adoption, and registration of regulatory documents and their upbringing to employees and organization of the employees' training.

A. Y. Makarov · O. Y. Antonova · R. I. Khabibullin
JSC "Bashkirian Power Grid Company", Ufa, Russia

A. S. Semerikov (✉)
Ural Federal University, Yekaterinburg, Russia

© Springer International Publishing AG, part of Springer Nature 2018
S. Syngellakis, C. Brebbia (eds.), *Challenges and Solutions in the Russian Energy Sector*, Innovation and Discovery in Russian Science and Engineering,
https://doi.org/10.1007/978-3-319-75702-5_9

Table 1 Consolidated plan of the development of regulatory documents

No.	RD type	RD title	Position of RD possessor	Position of the employee responsible for RD development	Ground for RD development	Scheduled date of RD development	Scheduled date of RD adoption
Branch of activities/functional areas							
1. . . .							

1.2 Transparency and Observability of Work Performance

The main objectives of the power grid company are high-quality, reliable power supply and well-timed technological connection to electricity networks. The company's customer-oriented approach requires creation of a transparent and open management system. For example, one of the first automated business processes was "management of material and technical support." Efficiency of the purchasing function acts as one of the conditions for uninterrupted process of production activity, capable of providing timely delivery of material and technical resources as well as execution of the corresponding work [1, 2].

1.3 Measurability of Work Performance Results

The plan of actions must ensure measurability of business processes performance at the enterprise. According to a widely spread opinion, it is possible to control such things that one is able to measure. If a business process becomes measurable, then it also becomes observable and controllable.

Besides, measurability of activity gives the possibility to carry out motivation of the branches' personnel based on objective rates. It became possible to allocate territorial subdivisions' rating and establish the order, amount, and conditions of employees' bonus awards [3].

Rating at the enterprise adheres to the following principles: data reliability (obtaining information on the basis of accounting reports, operating/technical reports on the ground of branch methods of indicators calculation); taking into account certain characteristics of the branch (geographic location, the customer basin, density of equipment service, amount of priority objects in the customer basin, remoteness of objects from the personnel's basing site and other characteristic features of outside environment); summation of the given indicators (calculation of estimated figures based on standard methods, their summation and indicators ranging on the principle of increase).

1.4 Adaptation to the Required Changes

An important quality of a modern management system is flexibility and well-timed adaptation to the changes of external environment. Due to processes speed ongoing both at the enterprise and outside, changes of legislative base, government institutions requirements, and schedule reduction, the company has to quickly and smoothly alter the business processes [4].

2 Procedures of the Enterprise Management System Improvement

The first stage of the program is "organizational and process transformations."

2.1 Optimization of the Organizational Structure

Based on the analysis of domestic and foreign practice of building up the enterprises management systems, we can speak about lack of uniform approach to the organizational structure. Even in case of a separately taken organization, it is impossible to speak about the selection of the best option because of continuous changes of environment and corporate processes [5]. In that context a list of effective solutions, regarding optimization of organizational structure, is suggested:

2.1.1 Determining Optimum Number of Management Levels

Three-level management system was chosen as the most efficient option for the power grid enterprise. Transition from a four-level management system to a three-level one lets us solve tasks concerning services quality improvement more quickly without reliability losses; thus, a power grid enterprise becomes more customer-oriented.

2.1.2 Development and Transition to the Standard Organizational Structure of the Branches

The organizational structure of the management company was created taking into account the principle of centralization of managerial and partially providing functions and with observance of economic feasibility. Formation of the standard organizational structure of territorial subdivisions allows one to increase efficiency of operating activities of the enterprise. This structure is variable and makes corresponding changes when the solution of additional tasks is required.

Transition to the standard organizational structure gave the instrument of comparison and the opportunity to create standard branches and, therefore, to record, analyze exceptions, and make decisions on them consciously.

2.1.3 Type Designs of Positions

Type design of positions means definition of a standard set of an employee's functional responsibilities, which excludes duplication of functions in the department, allows one to define a set of qualification requirements and to provide interchangeability, and, further, measures overall performance of employees.

2.2 Optimization and Regulation of Business Processes

Implementation of a full process approach in an electric grid company is not efficient. A more effective solution is the implementation of elements of process approach on the basis of functional organizational structure [5, 6]. The model of business processes at the top level was developed and approved in the company [1].

This gave the opportunity to determine problem areas and priorities of business processes optimization and regulation and carry out the following list of actions for each process:

1. Identification of the purposes and key performance indicators (KPI) of the business process
2. Modeling of the business process "as to be"
3. Formation of functional matrixes of the structural divisions participating in the business process
4. Synchronization of the business process and functional matrices
5. Development of the key performance indicators (KPI) of employees in the structural division with account of the business process KPI
6. Development of the indicators and reports for regular monitoring of the business process "health"
7. Formation and approval of regulations of business process

2.3 The Next Step of the Program Is Automation of Business Processes (ERP)

This stage is characterized by maintaining of business processes management system on the basis of information systems. Automation of business processes allows not only to transfer metering operation to IT environment but also to solve questions of horizontal interrelationship of the enterprise structural divisions.

Within the framework of the tasks stated above, the fundamental principle to change the approach to automation is applied at the enterprise. Automation implies the process from creation of the local unrelated programs to automation of previously optimized business processes in the unified information system (ERP) that provides automatic end-to-end integration of reference and operation data. According to the procedure approved by the company, any project must pass the following steps, which are precisely designated:

1. Choice of ERP system for automation of business processes (SAP, Oracle, 1C, etc.)
2. Choice of architecture and appointing of automation basic principles (the maximal use of standard patterns, centralized maintaining of reference data, end-to-end integration of operation data between configurations).
3. Test cases development for requirements elicitation for follow-up revision of standard patterns. Development of technical designs for each automated business process.
4. Standard patterns follow-up revision in accordance with the technical design and acceptance of alternations on the end-to-end test case.
5. Training of key users and creation of video instructions for mass teaching.
6. Monitoring of data input process in the course of the pilot operation.
7. Transfer to commercial operation.

Based on automation, the following projects (accounting and tax records, consolidation of International Financial Reporting Standards, etc.) are being successfully carried out at the enterprise (Fig. 1). The final stage implies the change of control and analysis system that includes two main task blocks (Fig. 2).

1. Automation of collecting and calculation of business process "health" indicators provides the company management with necessary current information for timely managerial decision-making and reduces labor input for its receiving.

   For example, for the automated material and technical support (MTS) process, including agreeing of contracts, the reports "optimization of material and technical support process" (Table 2) and "optimization of 'document flow: contracts' process" (Table 3) were worked out at the enterprise. These reports are placed at the enterprise portal that helped to increase transparency and observability of the process.

   The main stages that allow tracking the lot/contract are defined in the reports, thus acting as the effective instrument of internal control and assessment of the enterprise operation.

2. The purposes of automation of collecting and calculation of indicators of employees' efficiency in the structural divisions participating in business process are:

   • Monitoring of each employee's activity efficiency
   • Performing of internal comparative analysis of the staff performance [1]

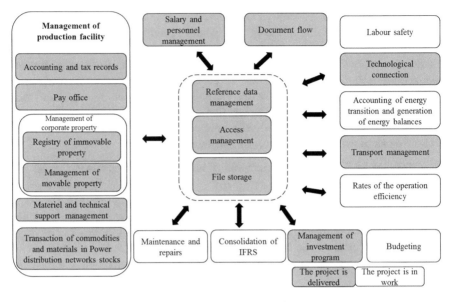

Fig. 1 Projects of automation

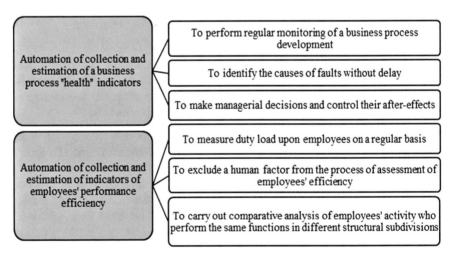

Fig. 2 The tasks of control and analysis system changes

The company realized the project to obtain the tool of collecting and calculation of 116 company employees' performance ratio (PR) through the example of "accounting and tax accounting" process. The system allows drawing a report, giving information on each employee (Table 3). Moreover, it is possible to produce individual PR calculation, thus giving an employee the opportunity to estimate his results and improve unsatisfactory indicators (Table 4).

Table 2 Optimization of material and technical support process

Stage.	Enterprise					
Lot, responsible for the lot	The term is not expired		Expired		Total	
Lot, lot title, number	Number of lots, pcs	Amount sum of lots, rbs	Number of lots, pcs	Amount sum of lots, rbs	Number of lots, pcs	Amount sum of lots, rbs
Stage 1						
Full name						
Lot 1						
Stage 2						

In prospect, the PR system will:

- Measure performance indicators of middle managers and entry-level employees.
- Carry out comparative analysis of activity of employees' performing the same functions in different departments when there is a standard structure of territorial branches.
- Exclude a human factor during data collection and PR evaluation (in the destination model it is planned to automate the process of data collection by 75–80%) (Table 5).

At the present moment, PR of material and technical support and services of technological connection to electricity networks are worked out, and they are under preparation for automation.

3 Conclusions

The following goals were reached after implementation of the program of the enterprise management system improvement: firstly, unified and standardized work performance on automated business processes; secondly, the condition of ensuring transparency and observability of work performance, which allowed to measure the results; and thirdly, the requirement of flexibility and adaptation of the system to outside environment changes in due time.

We can draw a conclusion that while implementing the comprehensive approach to the process of the management system improvement with the following steps: organizational and process transformation, automation of business processes, and changes of the monitoring system and the analysis, the program allows one to achieve the stated objectives.

As a result, a company's management can measure and monitor not only the timeliness and quality of each step of regulated and automated business process but also the job of every employee, who is involved in different business processes.

Table 3 Optimization of the process "document flow: contracts"

Stage			The term is not expired		Expired		Total	
Organization								
Document	Document prepared by	Document structural subdivision	Number of contracts, pcs	Amount sum of contracts, rbs	Number of contracts, pcs	Amount sum of contracts, rbs	Number of contracts, pcs	Amount sum of contracts, rbs
Stage 1								
Organization 1								
Contract no. 1								
Stage 2								

Table 4 The report of the subdivision PR

Employee, subdivision	PR № 1	PR № 2	PR № 3	PR № 4	PR № 5	PR № 6
Employee	Value	Value	Value	Value	Value	Value
Organization 1						
Subdivision 1						
Full name 1						

Table 5 Individual PR calculation

Employee				
The title of the indicator	Measuring unit	Working area 1	Working area 2	Totals
Full name				
PR 1				
PR 2				

References

1. Makarov, A.Y., Gurin, S.V., Antonova, O.Y.: Structured design in a power grid company. WIT Trans. Ecol. Environ. **190**, 37–44 (2014), Volume 1, WIT Press: UK
2. Nwankpa, J., Roumani, Y.: Understanding the link between organizational learning capability and ERP system usage. Comput. Hum. Behav. **33**, 224–234 (2014)
3. Rajan, C.A., Baral, R.: Adoption of ERP system: an empirical study of factors influencing the usage of ERP and its impact on end user. IIMB Manag. Rev. **27**, 105–117 (2015)
4. Aloini, D., Dulmin, R., Mininno, V.: Modelling and assessing ERP-project-risks: a Petri Net approach. Eur. J. Oper. Res. **220**, 484–495 (2012)
5. Makarov, A.Y., Gurin, S.V., Gorbachev, Y.V., Antonova, O.Y.: Evaluating a power grid company's employees' effectiveness. WIT Trans. Ecol. Environ. **190**, 119–126 (2014), Volume 1, WIT Press: UK
6. Repin, V.V., Eliferov, V.G.: Process Approach to Management, [in Russian], pp. 10–17. Mann, Ivanov and Feber, Moscow (2013)

Power Grid Infrastructure Modernization by the Implementation of Smart Grid Technologies

A. Y. Makarov, Y. A. Radygin, G. A. Sharafieva, and O. M. Rostik

1 Introduction

The power grid sector is one of the basic sectors providing the development of the national economy. This branch has a great impact on the development of other sectors and economy in general, and it means that ensuring reliable and high-quality power supply is an essential factor of economic and social stability.

The main questions faced by power grid companies, nowadays, are:

1. Current condition of grid infrastructure
2. Crucial problems demanding solution and expansion
3. Technical solutions that may be applied for current or future problems

The choice of appropriate technologies is the key aspect of giving the answers. Modern technologies of power grid infrastructure management are united by a catch-all term Smart Grid, the main elements of which are presented in Fig. 1. Various elements of Smart Grid technologies are relevant for different countries. These elements are determined by certain characteristics of power grid infrastructure of the given countries (Table 1).

Management of renewable energy sources that demands advanced technologies in regard to dispatching automated and remotely operated control is relevant for Europe. In the USA, much attention is paid to demand response, and smart metering prevails there. In BRICS countries, acute problems are outdated equipment and

A. Y. Makarov · Y. A. Radygin · G. A. Sharafieva
JSC "Bashkirian Power Grid Company", Ufa, Russia

O. M. Rostik (✉)
Ural Federal University, Yekaterinburg, Russia
e-mail: rostik@pm.convex.ru

© Springer International Publishing AG, part of Springer Nature 2018
S. Syngellakis, C. Brebbia (eds.), *Challenges and Solutions in the Russian Energy Sector*, Innovation and Discovery in Russian Science and Engineering,
https://doi.org/10.1007/978-3-319-75702-5_10

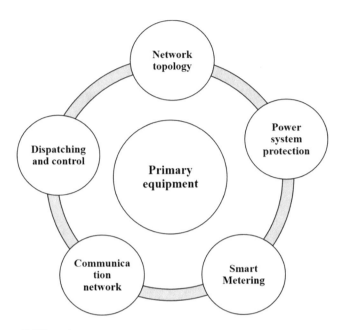

Fig. 1 Smart Grid key elements

Table 1 Characteristic features of Smart Grid technologies for different regions

Region	Characteristic features	Relevant Smart Grid technologies
Europe	Renewable sources Circular cable systems Widespread distributed generation CO_2 emission abatement	Medium and high voltage systems automation (hereafter MV and HV) GIS (geographic information systems), OMS (outage management systems), DMS (distribution management systems) Variable-delivery transformers for medium-voltage system Microgrids; smart metering
USA	Demand response Radial grids with overhead lines Unbalanced load Renewable sources	Microgrids; smart metering Demand response DMS, OMS, GIS High-voltage systems automation SCADA (supervisory control and data acquisition system)
BRICS countries	Outdated equipment Business losses Single-point large-volume generation	Medium- and high-voltage systems reconstruction and automation; SCADA for medium- and high-voltage sub-stations DMS, OMS, GIS Smart metering

energy losses; therefore, comprehensive grid modernization could be the method of solution.

Realization of one or another Smart Grid technology is designed to achieve the key purposes of power grid companies, in particular, to ensure [1]:

- Security of supply and energy quality
- Flexibility in the context of response to the network failures

To determine what Smart Grid means for the Russian Federation, it is necessary to thoroughly analyse the characteristic features of power grid infrastructure in Russian cities.

2 Performance Review of Distribution Systems of Large Cities in Russia and the CIS

A typical condition of city electrical grid in Russia and the CIS is described below through the example of the network in Ufa, Bashkortostan:

- Ufa's electricity supply is carried out by electricity networks of several voltage levels, 110/35/10/6/0.4 kV. Supply centres are 51 high-voltage substations (35, 110 kV). The overall number of 6–10 kV transformation stations is 2178.
- The network of 6–10 kV is mainly a simple open radial scheme, based on underground cables.
- Operational monitoring of power facilities parameters (power, currents, voltage levels) and switching off devices positions at substations of 35 kV and higher are implemented with an operational information complex (OIK – Dispatcher) in real-time regime. However, observability and remote control do not exist in 6–10 kV transformation stations and 0.4 kV network.

The described situation has significant drawbacks in comparison with the world's leading practices of the power grid management, and it is typical for the majority of large cities in Russia and the CIS. The average number of disconnected consumers and average outages duration in the network of "Bashkirenergo" exceed the corresponding figures in European countries [2] (Fig. 2). Alongside this, energy losses in Russian cities figure up to 10–15%, suggesting a significant potential for reduction (Fig. 3).

Benchmarking results show a significant gap between the operating efficiency of Russian and the CIS power grid companies and industry world leaders in the field of power grid infrastructure management.

Alongside this, depreciation of the equipment obsolescence also takes place. According to many positions average performance standards of the installed substation equipment of "Bashkirenergo" correspond to the equipment that was run in advanced countries 30 years ago.

To sum it up, a conclusion could be drawn that power grid infrastructure of big cities in Russia and the CIS has four characteristic system problems:

1. Poor reliability due to tough topology, which makes hard to identify a point of fault, fault condition spreads over large grid sections and loss in reliability due to cross-linking bonds

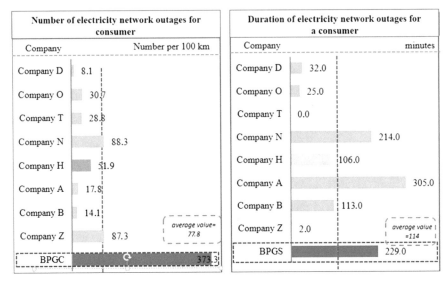

Fig. 2 Number of electricity network outages and outages duration as compared to European countries [3]

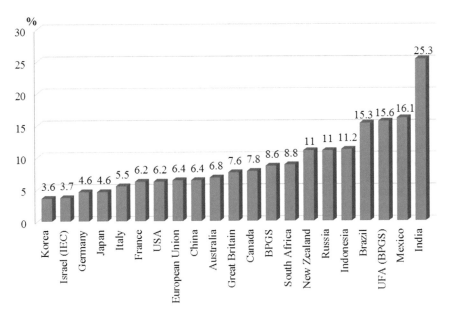

Fig. 3 Energy losses in different countries [4]

2. Poor manageability: absence of remote control and impossibility of control standardization
3. Relatively high energy losses and equipment depreciation

Taking into account the system problems stated above, the following elements of Smart Grid technologies are proposed to be relevant for the Russian Federation: network structure optimization by modern switching equipment; automation of network control, including its observability and remote control providing; automation of dispatching control; building the system of smart metering; and establishing cybersecurity of power facilities.

To determine Smart Grid technologies for future implementation, it is necessary to prepare a feasibility study for each city in review [2].

The work structure as defined within the feasibility study included:

1. Collection of initial data and modelling the city power grid
2. The analysis of grid current condition by applying the developed mode
3. Preparation of different scenarios of power grid perspective development, including the grid model and automation technologies offers for each scenario
4. Cost-benefit analysis for each scenario and selection of the preferred one
5. Plan preparation for the transition from the current situation to the power grid target model

The described approach is quite different from a traditional one applied for working out development programmes and schemes in the Russian Federation: to prepare the "as is" model, procedures both for providing projected increase in energy consumption and network topology optimization are offered. Moreover, this approach includes measures for grid automaton and power system protection improvement, for implementation of Smart Grid and connection systems.

In accordance with the results of feasibility study, "current power grid optimization and automation" and "building of electricity commercial metering system" measures turned out to be the best in the context of development prospects of Ufa's power grid infrastructure.

3 Comprehensive Modernization Project of Ufa Power Grid Infrastructure

Within the framework of a feasibility study, a comprehensive plan of transition from the current situation to the target model with an activity progress chart and detailed estimation of capital investments required was worked out. The transition plan is intended for 5 years and includes the following stages: reconstruction and automation of 512 substations including observability and remote control and reconstruction of Ufa power grid management centre with implementation of SCADA-system, all related systems and 80,000 metering instruments setup.

The calculations showed that upon the transition from the current situation to the target model, the following effects will be reached:

• Business losses reduction by 80% from the current level
• Technical losses reduction by 30% from the current level

- Reduction of power outages duration owing to reduction of failures by 50%
- Economy of time for faults searching and switching by 70% due to grid observability and structure optimization
- Current equipment service life extension by 10%

That sort of a project is being implemented in Russia for the first time. Hence, to choose the best technical solution for the main project tailored for 5 years, JSC "BPGC" made a decision to execute a pilot project in Ufa in the following steps:

- Reconstruction of two distribution and five transforming substations of "Bashkirenergo" Ufa's East Region Electric Power Grids
- Construction of the network management centre building (hereafter NMC)
- Creation of the automatic dispatch system of the network management centre, including PSI and multiple depiction system.

The sketch map of the pilot region network section and the entire Smart Grid project is presented in Fig. 4. The pilot project is being executed by "BPGC Engineering" – affiliated company of JSC "BPGC." During the pilot project execution, JSC "BPGC" specialists worked out an innovative approach to the grid automation that provides complex observability and manageability while reconstructing only 25% of equipment.

Based on the data collected during the pilot project implementation, preparation for the main stage of the project is ongoing at the moment – Ufa power grid reconstruction is being designed. Also, it is necessary to point out that a number of problems occurred during execution of the pilot project.

Due to different approaches to grids manageability in the Russian Federation and western countries, difficulties in direct adaptation of western solutions and necessity to consider specific features of Russian cities' network structure occurred while performing the feasibility study. It was essential to establish technical solutions for further extension at the stage of design and survey work. While addressing the issue, the decision was made to focus on modern standards instead of traditional solutions of automated technological process control system concerning data communication protocol, remote supervision and surveillance. Protocol 61,850 was chosen as the data communication protocol.

The following difficulties arose at the stage of the pilot project execution: dependence on the equipment manufacturer on the part of production time, delayed delivery of switchgears, difficulties in equipment installation due to its complexity and division of responsibilities between the customer and the supplier and significant cost overrun. As a solution, it was decided to launch own manufacturing of the power network equipment to downplay the problems stated above.

For implementation of Ufa power grid comprehensive modernization project with Smart Grid technologies applied, an assembly of smart distribution devices based on "Siemens" switching equipment started up on the territory of the Republic of Bashkortostan by virtue of the agreement between "Siemens" LLC and JSC "BPGC." All secondary circuits are assembled by JSC "BPGC."

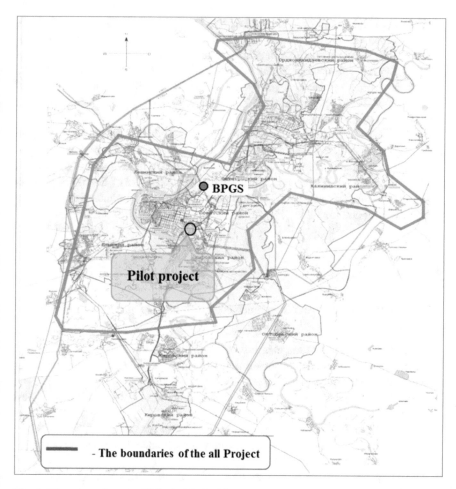

Fig. 4 The schematic map of Ufa electrical grid

Table 2 The analysis of price relation of various electrical switchgear producers

Technically identical electrical switchgear	Price relation, ea
European analogue	3.2
European equivalent	1.5
Domestic equivalent	1.3
Domestic equivalent	1.1
Gas insulated switchgear 8DJH Siemens 1250	**1.0**
Domestic equivalent	0.8

A pricing policy review showed that the gas insulated switchgear 8DJH by JSC "BPGC" and "Siemens" competes with the equipment of low-price segment regarding its price and with expensive devices one regarding its quality (Table 2).

Thanks to the usage of component parts by different producers, the balance between functional characteristics, price and quality of electrical switchgears relevant for power grid reconstruction projects was reached.

JSC "BPGC" will be able to reduce the volume of investments required for the project by setting up its own production. Besides, it will establish a centre of excellence in the Republic for further distribution of Smart Grid technologies in other regions of the Russian Federation.

JSC "BPGC" is the first power grid company in Russia implementing Smart Grid technologies. Best practices of JSC "BPGC" in implementation of power grid comprehensive modernization with Smart Grid technologies and production of modern electrical equipment necessary for it in the Republic of Bashkortostan are highly important and have great opportunities for power grid modernization in the cities of Russia and the CIS.

4   Conclusion

Based on the pilot project implemented in Bashkirenergo, the following are concluded:

1. Smart Grid implementation is impossible without a preliminary feasibility study.
2. Technologies providing observability and manageability are preferable for Russian power grids.
3. To ensure successful implementation of the technologies, it is reasonable to test them within a pilot project framework.
4. To provide sufficient speed and flexibility of the order execution, the company performing the comprehensive reconstruction must cooperate with the manufacturer of the equipment in use.
5. The company implementing Smart Grid must develop skills sets in the sphere of checkout of this equipment in order to be able to reconstruct it more flexible and independent from the contractor.

References

1. European Commission Directorate-General for Research Information and Communication Unit European Communities: European Technology Platform Smart Grids, Vision and Strategy for Europe's Electricity Networks of the future, European Communities (2006)
2. Makarov, A.Y., Hain, Y., Bayramov, I.Y., Podshivalov, V., Radygin, Y.A., Koifman, E., Krutous, I.S.: The comprehensive modernization of Ufa's electricity infrastructure. WIT Trans. Ecol. Environ. **190**, 559–569 (2014). Volume 2, WIT Press: UK
3. Potential assessment of JSC "Bashkirenergo" power grid efficiency upgrading. McKinsey & Company, 2011, 162 p.
4. Smart Grids: Best Practice Fundamentals for a Modern Energy System, p. 6. World Energy Council Review, London (2012), www.worldenergy.org

Customer Orientation of Electricity Retail Sales Companies

B. A. Bokarev, D. G. Sandler, and M. V. Kozhevnikov

1 Introduction

There is a fairly large number of negative factors that affect electricity retailers, including a general trend among consumers to delay payments, major industrial consumers moving away from last resort suppliers, a lack of a clear strategy on the part of the government as to how to develop electricity markets as regards tariff regulation, energy conservation and legal provisions.

Evidence from many countries that had their electric power sector restructured shows that regardless of the environment that electricity suppliers operate in, they lose their monopoly-induced competitive advantages as the electricity and capacity markets develop. The competitive advantages of electricity retailers can only stem from their intrinsic potential that consists of their:

- Social capital
- Key domain-specific competencies
- *Focus on current and potential needs of customers*
- A management system that makes it possible to maintain and develop the above-mentioned factors

There is, therefore, a need to create a methodological platform for implementing a customer-oriented approach to ensure the sustainable development of electric power companies and, more specifically, retail sales companies.

B. A. Bokarev
JSC "Atomenergosbyt", Moscow, Russia

D. G. Sandler · M. V. Kozhevnikov (✉)
Ural Federal University, Yekaterinburg, Russia
e-mail: m.v.kozhevnikov@urfu.ru

© Springer International Publishing AG, part of Springer Nature 2018
S. Syngellakis, C. Brebbia (eds.), *Challenges and Solutions in the Russian Energy Sector*, Innovation and Discovery in Russian Science and Engineering,
https://doi.org/10.1007/978-3-319-75702-5_11

2 Customer Orientation Issues in Electric Power Industries

The key principles of customer orientation can be outlined as follows:

1. A customer (or group of customers) is selected within the Company's management system.
2. The Company provides a targeted response to the customer's demands and clearly expressed needs.
3. The customer is evaluated and ranked on the criteria of revenue and other specific characteristics that matter for the achievement of the Company's strategic goals.
4. Effective projects and measures are developed and implemented in a systemic way to manage the satisfaction of the customer's known demands and clearly manifested needs.
5. The perceived and created customer value of the Company's products is studied.
6. Systemic projects are implemented that are aimed at managing the created and perceived value of the Company's products for the customer. The characteristics and range of products and services are adjusted, and the operational functions of the business are re-engineered accordingly.
7. The Company engages customers in value chain management.

A customer-focused approach implies that pinpointing and satisfying the needs of customers whom the company "knows personally" is the only way to ensure its leadership in the industry, to make its products competitive and, ultimately, to meet the objectives set by its owners and reach various (financial, social and other) goals of the company. The electricity sectors of developing countries remain highly monopolized despite reforms, and consumers are not given priority as the prime factor of business development.

An energy company can opt for one of two tariff policy options that differ critically in terms of customer relations:

1. The strategy of short-run profit maximization aimed at gaining the maximum yield from customers "today"
2. The strategy of long-run profit maximization aimed at encouraging effective demand, i.e. at supporting and building up a customer base as the only source of income

The advantages and disadvantages of each strategy are outlined in Table 1.

The majority of Russian electricity companies are now implementing the first strategy. Among other things, this results in manufacturing companies losing their competitive edge over similar products imported from abroad because their energy costs become extremely high, while energy suppliers do not make any efforts to reduce energy intensity [1]. Having one's own power generation facilities becomes the only way-out, which naturally leads to lower revenue for energy companies. This creates a vicious circle in which consumers and, eventually, energy companies themselves are faced with growing prices of goods and services (Fig. 1). Consequently, products generated by the power system become less competitive.

Table 1 Advantages and disadvantages of different tariff strategies

Strategy	Advantages	Disadvantages
Short-run profit maximization	• Excess profit made quickly in the short run • No need to invest in R&D and innovation	Passing on costs to consumers through higher tariffs could lead to: • Lower electricity consumption because of customers converting to alternative energy sources or, when possible, leaving for other electricity suppliers • Non-payment or delayed payment of bills • Unauthorized connections to the grid and resulting losses
Long-run profit maximization	• Stable demand for electricity and growing revenue of the electricity supplier • Social and economic benefits for the state, local government and society as a whole	• Government and financial institutions have to be involved in developing investment and energy programmes

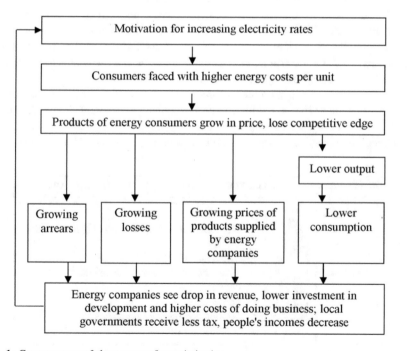

Fig. 1 Consequences of short-run profit maximization

3 Methodology for Assessment of Customer-Oriented Management Approach

Figure 2 depicts the customer structure of an energy retailer. It is worth noting that all too often some business partners whose interests the company seeks to satisfy are mistakenly excluded from the ranks of its customers because they do not consume its key product/service.

To assess the customer-oriented management approach, we shall use a system that will be drawn as a radar chart where the number of axes represents the number of customers (or groups of customers) of the Company and the scale on the graph characterizes the strength of the Company's focus on a specific customer (or group of customers) by means of anchored data markers with detailed labels. The number of axes for the chart is selected specifically for each business and might represent either the list of objects of focus (Fig. 3a) or an expanded group of identical objects (Fig. 3b).

We refer to the customer mix of the Company as its orientation field. The resulting star chart that illustrates the expert evaluation of the Company on the suggested scale characterizes the level of orientation.

We shall refer to the gap between the current and target level of customer orientation of the Company (as envisaged in its development strategy) as its orientation potential. The size and shape of the orientation potential in the chosen orientation field will determine the choice of the ways and means for implementing the customer-focused management approach.

The frequency of adjusting the customer orientation of the system, as indicated in the classification in Fig. 3, might be one-off, regular and constant. If it is enough for the Company to adjust its customer focus once, in most cases, there will be no need for changing the business structure or business process re-engineering. If orientation

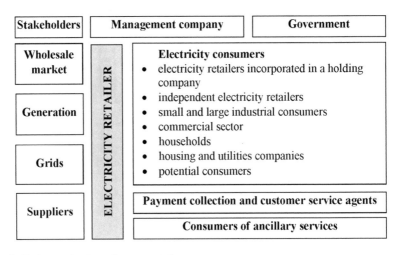

Fig. 2 Customer structure of energy retailer

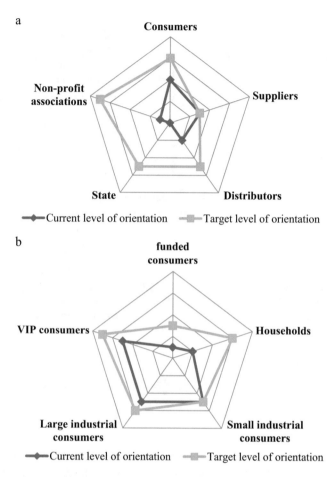

Fig. 3 Example of comprehensive assessment of customer-focused management approach. (**a**) Based on full list of objects of focus. (**b**) Based on expanded group of identical objects

adjustment has to happen regularly or constantly, the Company cannot avoid redesigning its management system and re-engineering its business processes appropriately [2].

Modern information technologies make it much cheaper to launch the orientation process once and keep it up later on, rather than carry out periodic surveys which require a lot of resources. We would like to emphasize that keeping the customer base constantly up to date later on is not only cheaper than periodic surveys but is also less annoying to customers because it occurs unobtrusively in the course of routine transactions.

The next step is to choose a management style of the two available options: management by disturbance and management by exception. Management by disturbance, unlike management by exception, stipulates that the system should be adjusted as soon as changes in the customer mix or individual customer

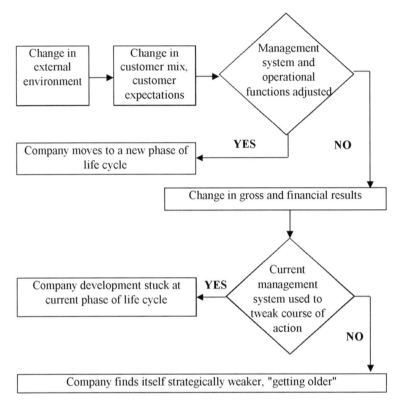

Fig. 4 Advantages of managing customer orientation by disturbance

characteristics are revealed. For the management approach to work, the Company needs to be able to perform monitoring and forecasting, as well as to design control activities within a limited timeframe and respond to them (Fig. 4).

Management by disturbance is more difficult to arrange and costlier to employ than management by exception. It is, therefore, necessary to decide whether it will prove effective for managing customer focus with regard to a certain customer group. In summary, to evaluate the customer-oriented approach used by a company, one should:

- Define the orientation field of the Company, i.e. the customer mix that the Company focuses its business on in order to achieve its strategic goals.
- Establish the current and set the target level of customer orientation for each customer (or group of customers) in the orientation field, thus identifying the Company's orientation potential.
- Determine the frequency of orientation for various groups of customers.
- Identify the current approach to customer-oriented management and set the target one for each group of customers and decide whether management by exception is sufficient or it should be replaced with management by disturbance.

- Map out the current state of customer-oriented interaction and set the target state of customer-oriented interaction, including factors that the company would want to modify when adopting proactive management practices.

4 Differential Electricity Pricing

Dynamic pricing is an example of a customer-oriented approach used by electricity retailers. Dynamic tariffs are offered to consumers who agree to supply interruptions during times of peak demand in return for monetary benefits in the form of discounts. The size of the discount depends on the number of interruptions during the billing period, their duration and the volume of undelivered electricity (capacity) per break. The consumer chooses the desired parameters from a "menu" proposed by the energy company.

Dynamic tariffs are, therefore, designed to reduce systemic costs of electric power companies and lower electricity bills for consumers by increasing prices during peak hours and decreasing them during off-peak hours [3, 4].

Applying various tariffs has different impacts on the level of economic risk borne by energy consumers and utilities. For example, from the point of view of consumers, traditional fixed tariffs are the least risky because the rates do not depend either on the amount of electricity consumed or the time of consumption. On the contrary, such tariffs are extremely inconvenient to the utility because during peak hours it produces electricity at a loss. Figure 5 illustrates how the risk borne by energy market participants depends on tariffs.

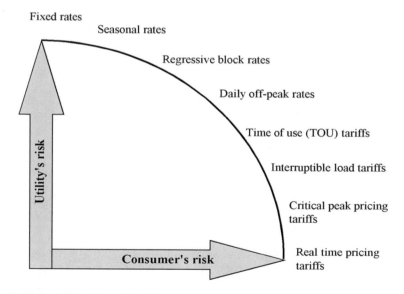

Fig. 5 Markets' risks of the utility depending on the tariff in effect

5   Conclusion

The proposed system for evaluating a customer-oriented management approach makes it possible to clearly and explicitly set objectives for a course of action aimed at transforming the management system in an electricity retail sales company, to build a marketing concept, to structure the current marketing activity integrated into the business processes of the company and to make decisions with regard to its product policy and modify its tariff menu. The combination of these makes the utility more competitive and solidifies its market positions.

Acknowledgement   The work was supported by Act 211 Government of the Russian Federation, contract № 02.A03.21.0006.

References

1. Karpovich, A.I.: DSM for economic soundness of energy companies, [in Russian]. National Interests Priorities Secur. **22**, 2–5 (2011)
2. Bokarev, B.A.: Concept of Customer-Oriented Management, [in Russian], Proc. of the 8th Int. Conf. on Public Administration in the 21st Century: Traditions and Innovations. University Press, Moscow (2010)
3. Cousins, J.T.: Using time of use tariffs in industrial, commercial and residential applications effectively.   http://www.tlc.co.za/white_papers/pdf/using_time_of_use_tariffs_in_industrial_commercial_and_residential_applications_effectively.pdf
4. Gitelman, L.D., Ratnikov, B.E., Kozhevnikov, M.V.: Demand-side management for energy in the region. Economy Region. **2**, 71–78 (2013)

Methodological Approach to Energy Consumption Management at Industrial Enterprises

A. P. Dzyuba, I. A. Baev, I. A. Solovieva, and L. M. Gitelman

1   Introduction

In conditions of globalization and increased international competition, the issue of energy efficiency increase is one of the most vital in the world economy [1]. The task of energy efficiency increase is particularly important in the industrial sector, whose share of energy consumption in the total energy balance of most energy-intensive countries of the world, such as Russia, Finland, and Germany, considerably exceeds worldwide average indices (Fig. 1).

Most research devoted to industrial energy efficiency increase is focused on obtaining a result in the production and technical environment of enterprises. At the same time, a significant share of reserves of reduction in industrial energy consumption remains in the environment of energy consumption planning, organization, and control. The task of development of methods and tools for energy efficiency increase in the planning environment has become most vital after the development of energy market mechanisms [4]. The main and, at the same time, the less studied line of energy efficiency increase is the development of methods and mechanisms based on the tools for forecasting of energy consumption parameters.

The cost of electric energy purchased by industrial enterprises in the energy market consists of several components, the basic ones being electric energy and electric power. All industrial enterprises also pay for services tradable outside market

A. P. Dzyuba
LLC Chelyabinsk Office of Power Trading, Chelyabinsk, Russia

I. A. Baev · I. A. Solovieva
Department of Economics and Finance, South Ural State University, Chelyabinsk, Russia

L. M. Gitelman (✉)
Department of Energy and Industrial Management Systems, Ural Federal University, Yekaterinburg, Russia

© Springer International Publishing AG, part of Springer Nature 2018
S. Syngellakis, C. Brebbia (eds.), *Challenges and Solutions in the Russian Energy Sector*, Innovation and Discovery in Russian Science and Engineering,
https://doi.org/10.1007/978-3-319-75702-5_12

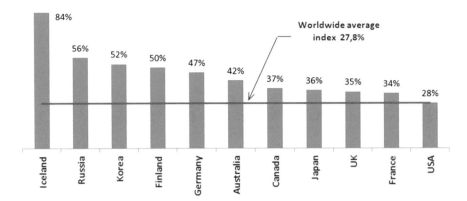

Fig. 1 Share of industrial energy consumption in several countries [2, 3]

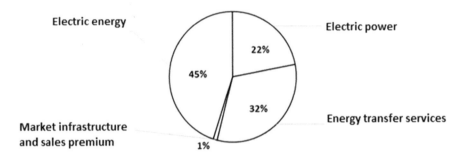

Fig. 2 Structure of costs on industrial enterprises' purchase of electric energy in the Russian energy market

relations – the service on transfer and provision of the market infrastructure, whose tariffs are regulated by the state (Fig. 2). It should be noted that the tools offered can also be used by industrial enterprises of different countries, as the energy market model of Russia is developed on the basis of energy markets functioning worldwide [5, 6].

2 Research Methodology

Based on the electric energy cost components, as well as pricing characteristics in each component, we offered an overall structure of methodological approach to energy consumption cost planning (Fig. 3). The basis of the model is division of energy consumption components into electric energy and electric power, as well as division of energy consumption impact environments into process and market [7].

The electric energy cost planning model is based on the two-level mechanism grounded in short-term forecasting of the industrial enterprise energy consumption

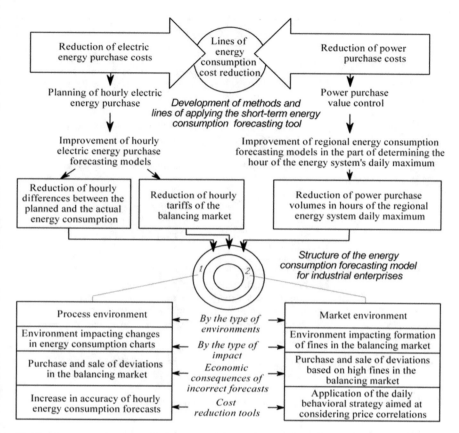

Fig. 3 Structure of the methodological approach to industrial enterprises energy consumption cost planning

parameters and considering the market environment factors (Fig. 4). The electric energy price consists of two components – the price of planned volume ordered by the enterprise for purchase one day before the actual electric energy delivery date and the price of deviations of the actual energy consumption volume from the planned one. Incorrect forecasting of hourly electric energy purchasing plans leads to appreciation of the electric energy cost in the form of fines for each error [8].

However, even at the highest accuracy of hourly energy consumption planning, due to the impact of many factors, errors in plans and, consequently, fines are inevitable. In this connection, during purchase of electric energy in the energy market, in our opinion, it is necessary to forecast and account for price factors forming fines or, in other words, market environment factors, which will allow one to reduce fines (see Fig. 3).

We propose to consider the market environment factors by forecasting the most favorable directions of deviations between planned and actual hourly energy

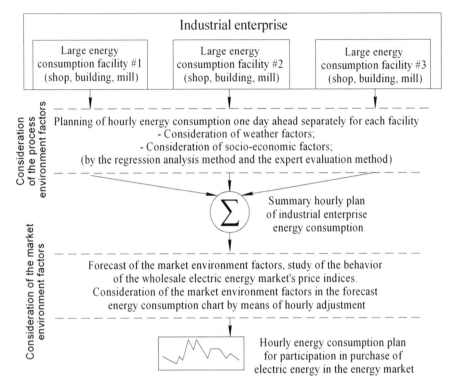

Fig. 4 Structure of the electric energy purchase cost planning model

consumption volumes and further adjustment of hourly plans in the knowingly favorable direction in terms of fines minimization [7].

The amounts of hourly fines for deviations in the balancing market are calculated as the difference between the price of the day-ahead market (PDAM) and the price of the balancing market (PBM) (Eqs. (1) and (2)). This is preconditioned by the fact that the market participant in any case purchases electric energy at the scheduled energy consumption price – PDAM, therefore, expenses incurred above the planned value prices are taken as fine amounts (Eqs. (3) and (4)).

Deviation purchase price:

$$P_{BM} \ \text{purchase} = \max \ (P_{DAM}; P_{BM}), \tag{1}$$

deviation sales price:

$$P_{BM} \ \text{sale} = \min(P_{DAM}; P_{BM}), \tag{2}$$

where P_{BM} purchase – purchase price in the balancing market; P_{BM} sale – sales price in the balancing market; P_{DAM} – price of the day-ahead market; P_{BM} – price of the balancing market.

Table 1 Combinations of different correlations of prices and volumes in the balancing market

№	Correlations		Market transaction	Calculation formula	Transaction price	Wholesale market participants' behavior strategies
	Price	Volume				
1	$P_{BM} > P_{DAM}$	Vplan > Vact	Sale	Min (P_{DAM}; P_{BM})	P_{DAM}	Sale of excessively purchased volume at a lower price. Fine in the form of price difference
2	$P_{BM} > P_{DAM}$	Vplan < Vact	Purchase	Max (P_{DAM}; P_{BM})	P_{BM}	Purchase of deficient volumes at the same price as planned volumes. No fines
3	$P_{BM} < P_{DAM}$	Vplan > Vact	Sale	Min (P_{DAM}; P_{BM})	P_{BM}	Sale of excessive volumes at the same price as planned volumes. No fines
4	$P_{BM} < P_{DAM}$	Vplan < Vact	Sale	Max (P_{DAM}; P_{BM})	P_{DAM}	Purchase of deficient volumes at a higher price. Fine in the form of price difference

Fine amount upon purchase:

$$P_{BM}\text{fine}_{\text{upon purchase hour}} = \max(P_{BM}; P_{DAM}) - P_{DAM}, \qquad (3)$$

fine amount upon sale:

$$P_{BM \text{ fine upon sale hour}} = |\min(P_{BM}; P_{DAM}) - P_{DAM}|, \qquad (4)$$

where $P_{BM \text{ fine upon purchase hour}}$/$P_{BM \text{ fine upon sale hour}}$ – fine amount upon electric energy purchase and sale in the balancing market for one specific hour (rub/MWh).

Analysis of the variants of directions of price relations in the day-ahead market and the balancing market and mutual deviations of planned and actual energy consumption volumes allowed one to outline combinations, upon occurrence of which market participants do not incur losses in the part of payment of fines in the balancing market, in spite of inconsistencies between forecast and actual hourly energy consumption values (Table 1). The outlined correlations formed the basis of an energy consumption forecasting model and formation of electric energy purchase bids in the energy market.

Then, we will consider in detail the approach to electric power purchase cost planning. The amount of liabilities is calculated for each month and determined as an arithmetical average of electric power consumption values in the hour of the daily maximum of working days of the electric energy system of the region, where the industrial enterprise purchases electric power (Eq. (5)):

$$\bar{P}_{month} = \frac{\sum\limits_{i=work.days}^{n} (P_i \in t_{maxi})}{n_{work.}} \qquad (5)$$

where $\bar{P}_{month}$ is the amount of liabilities on purchase of power (MW·month.), n_{work} is the number of working days in the estimated month, P_i is the amount of the industrial enterprise's power in the hour of the daily maximum of the region's electric energy system (MW), and t_{maxi} is the number of hour, when the daily maximum of the region's electric energy system is formed.

3 Practical Application of the Considered Approaches

In the course of application of the proposed mechanism at the industrial enterprise, the main task is forecasting of the hour of the daily load maximum of the electric energy system of the region, where electric power is purchased. An example of diagrams of shares of distribution of hours of the energy systems' load maximum for two regions is presented in Fig. 5.

As a result of empirical analysis of hourly daily charts of electric energy system load of Russian regions, we have outlined and generalized factors impacting the formation of the daily maximum's hour (Table 2).

Taking into account the outlined factors in the formalized model will allow one to forecast the daily maximum's hour for the energy system of any region with a sufficient degree of probability.

The obtained forecast shall be further considered while organizing the industrial enterprise production processes, which, in its turn, will allow the power purchase expenses to be cut down.

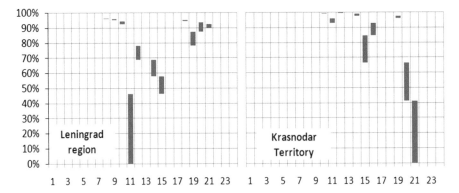

Fig. 5 Diagrams of shares of distribution of hours of the energy system energy consumption maximum during working days of 2013 (Moscow time)

Table 2 Summary table reflecting the impact of factors on parameters of the hourly daily chart of the regional energy system load

| Impacting factors | Factor sphere of impact | | | | Degree of impact on the peak load maximum hour |
	Total energy consumption value	Hourly chart form	Morning peak form	Evening peak form	
Climate and geographical conditions of the region location	+	+	+	+	High
Time zone		+	+	+	Average
Planned hours of the energy system peak load			+	+	Below average
Level of the region socioeconomic development	+	+	+	+	High
Share of energy consumption by the region industry	+	+	+		Average
Length of the light day		+	+	+	Average
Season	+	+	+	+	High
Month	+	+	+	+	High
Type of the week working day		+	+	+	Average
Individual features of the region energy consumption	+	+	+	+	High
Climatic changes in the region	+	+	+	+	High

4 Conclusion

The methodological approach developed allows management of energy expenditures on purchase of both electric energy and power in an integrated manner. Planning of purchasing costs of electric energy does not only increase forecast accuracy but also accounts for the market environment factors, which allows one to significantly reduce losses preconditioned by the energy market penal sanctions even in the presence of errors in hourly plans. In planning power purchase costs, the model proposed also accounts for the regional energy system daily maximum hour, which allows one to considerably reduce the amount of power payment liabilities at an insignificant change of production process charts.

The practical importance of the methodological approach lies in its universality and applicability to different types of industrial enterprises, irrespective of energy consumption volumes, field characteristics of load patterns, as well as regions purchasing electric energy [9]. The practical approval of the model at industrial enterprises underlined its ability to reduce the number of hourly plan errors by up to

5%, to cut down expenses on fines connected with outlining the market environment factors by 10%, as well as to lower power payment liabilities by 10–50%. Reduction of the total energy consumption cost due to using the complex of proposed methods ranges from 2% to 20%.

References

1. Key world energy statistics 2014; International Energy Agency. www.iea.org/books
2. Functioning and Development of the Electrical Energy Industry in the Russian Federation in 2011, Informational and analytical report of the RF Ministry of Energy; Agency for Electrical Energy Balance Forecasting CJSC (2012). http://minenergo.gov.ru/upload/
3. Energy in Sweden 2010, Facts and figures. https://www.energimyndigheten.se
4. Industrial energy efficiency: Using new technologies to reduce energy use in industry and manufacturing; Environmental and Energy Study Institute. http://ladoma.org/
5. Stridbaek, U.: Lessons from Liberalised Electricity Markets, 222 p. OECD/IEA, Paris (2005)
6. The Nordic electricity exchange and the Nordic model for a liberalized electricity market. http://nordpoolspot.com/globalassets/download-center/rules-and-regulations
7. Baev, I.A., Solovieva, I.A., Dzyuba, A.P.: Forecasting of industrial energy consumption in conditions of price signal volatility [in Russian]. Regional Economy. **4**, 109–116 (2012)
8. Gitelman, L.D., Ratnikov, B.E.: Economics and Business in the Electrical Energy Industry: Cross-Disciplinary Manual. Economics, Moscow (2013)
9. Renewing Ontario's electricity distribution sector: putting the consumer first. Message from the Chair; Website of the Ministry of Energy Ontario. http://www.energy.gov.on.ca/en/ldc-panel/

Part III
Investment Mechanisms for Energy

Model Investment Mechanism of Cost Management in the Oil Business Based on *RAROC* Methodology

A. Domnikov, P. Khomenko, and M. Khodorovsky

1 Introduction

The oil and gas business holds a leading position in the world economy and is the basis of international economic integration and formation of investment potential. Oil and gas projects related to the exploration, production, transportation, and oil facilities – refineries, petrochemical plants, and pipelines – are costly and have a long payback period (more than 10 years). The adequacy of investment from the preparation of the resource base and creation of new facilities to repair and reconstruction of existing facilities is the basis of normal reproductive processes in the industry. Therefore, there is a problem of making investment decisions. Correct investment decisions will ensure the sustainable development of the business, strengthening the competitive positions and increasing business value. An incorrect decision will result in the loss of market share, the loss of capital, and the destruction of value. The most important element of the mechanism of making investment decisions is the methodological apparatus of the risk assessment and methods of making and minimizing.

In addition to the definition of objects and volumes of investment, the problem of making investment decisions is the reverse side associated with the capital raising and financing in the oil and gas industry. The need to ensure long-term financial flows increases the importance of managing the strategic stability of an oil and gas company, determined by the target credit ratings and the target cost of funding, margin *ROIC*, *EVA*, and other related indicators. In this regard, the evaluation mechanism and the management of investment risk should be focused on achieving long-term sustainability and contribute to value creation in the long term.

A. Domnikov (✉) · P. Khomenko · M. Khodorovsky
Department of Banking and Investment Management, Ural Federal University, Yekaterinburg, Russia

© Springer International Publishing AG, part of Springer Nature 2018
S. Syngellakis, C. Brebbia (eds.), *Challenges and Solutions in the Russian Energy Sector*, Innovation and Discovery in Russian Science and Engineering,
https://doi.org/10.1007/978-3-319-75702-5_13

2 The Cost of Business, Investment Risk, and Financial Decisions

The cost of business in modern management practices is a key comprehensive indicator of business performance. The *EVA* concept, developed in the 1980s of XX century, became widespread in the modern economy and is embedded in the management systems of major corporations around the world. This concept is based on the idea of Adam Smith that the investment of capital should be created at a minimum required rate of return. This required rate of return applies to loan capital and to the owned one. In the general case, *EVA* is defined as

$$EVA = (ROIC - WACC) \times IC, \tag{1}$$

where *EVA* is the economic value added, *WACC* is the weighted average cost of capital, and *IC* is the amount of capital invested.

A positive value of *EVA* characterizes the efficient use of capital. Investors profit if the value of *EVA* is equal to zero. A negative *EVA* value characterizes inefficient use of capital.

The driver of value creation is spread efficiency, defined by the difference between *ROIC* and *WACC*. This spread is a key indicator of ranking the business units and implemented investment projects for the created value.

At the end of XX century a risk-based approach to managing the business value is spreading. Under this approach, the risk-adjusted return on capital (*RAROC*), a risk-adapted system of assessment and management of business performance through the use of economic capital model, is the most popular [1]. The system allows one to assess and compare the economic profitability at the level of the bank as a whole and at the level of specific transactions and business units, which have different levels of risk. If you use other assessment indicators, such as profitability, ROA/ROE, this is not possible, because the results obtained (e.g., rate of return) do not reflect the amount of risk of a transaction or business unit, which achieved this result. The *RAROC* calculation is carried out according to the following formula (2):

$$RAROC = \frac{NI - EL}{ECAP}, \tag{2}$$

where *NI* is net income, *EL* is expected loss (determined by the amount of regulatory provisions), and *ECAP* is economic capital.

As part of this approach measure the business value is transformed on the basis of profitability, adjusted for the level of risk taken:

$$EP = (RAROC - HR) \times ECAP, \tag{3}$$

where *EP* is the economic profit, which characterizes the value added business, and *HR* is the hurdle rate, which characterizes the required return on equity, estimated through the model CAPM.

Given the transformation *RAROC* in economic benefits, it is easy to show which investments create value and which destroy it by comparing the *RAROC* and *HR*. If the value exceeds the *RAROC* hurdle rate, the investment creates value, if *RAROC* < *HR*, the price collapses [2].

Risk assessment of investment projects and the calculation of *RAROC* components will be discussed later.

3 The Basic Components of the Model Economic Capital

The main parameters that characterize an investment project in order to assess economic capital are [3–5]:

- *PD* (probability of default) – the probability of default is the main indicator of the level of risk of the project and reflects the likelihood of default on the investment project.
- LGD (loss given default) – the level of loss given default is the expected average relative size of the losses of the company upon default of the investment project. This part of the cost of the investment project will be free of charge lost in the case of default. The introduction of the measure is justified by the fact that in the case of default of the investment project, it can be implemented fully or partially by way of sale, billing insurance requirements and options and other ways.
- EAD (exposure at default) – the position at risk characterizes the absolute value of the sum of the investment project and determines its full actual or forecasted cost of the investment, operating, and other costs.
- M (maturity) is the effective date. This is the average time during which a stored position on risk is determined by the term of the investment phase of the project.

The Long-term investment phase of the projects will lead to increased risks; the consequence of greater uncertainty results from the implementation. Less long-term investment phase reduces the overall risk of the project.

The evaluation model for *PD* is part of the calculation of risk capital described earlier. Economic capital is calculated taking into account the probability of default of the investment project. Parameter estimate LGD is obtained in four stages:

- Preparing data for modeling. This is based on the statistics of implementation of investment projects of the company for a long period of time (not less than 3 years). The main evaluation parameters can be LGD for each project, published in a default, or RR (recovery rate) – the rate of return that characterizes a fraction of the cost of the project, returning in the form of cash flow after the default.
- Classification of investment projects according to the criterion of significance of differences of the value of LGD. The selection of homogeneous groups of

investment projects may be selected based on the criteria of scope, objectives, timing, type of effect, deadlines, types of cash flow, the state of the economy, and other criteria. The final grouping is done based on the criterion of significance of differences of averages in a sample that can be evaluated on the basis of a t-test, F-test, Fisher, Kolmogorov-Smirnov test, and U-Mann-Whitney test [6].

- Build the LGD distribution for selected groups. Based on data from statistics, LGD is the built distribution for each classification group.
- The evaluation of the shape of the distribution of LGD and determination of main parameters. At this stage, the evaluation of the shape of the distribution defines the parameters for modeling LGD for each classification group. The evaluation of the shape of the distribution may be carried out using criteria χ^2, Anderson-Darling, and Kolmogorov-Smirnov [6].

In modeling economic capital, model Merton-Vasicek will use numerical values of LGD. However, the calculation of risk capital by simulation of LGD can be used as a random variable with the specified parameters defined in stage 4 described above.

The effective term describes a fine on the long-term investment phase. Further correction in risk capital for the project for more than 1 year is carried out according to the formula (4) [7].

$$M = \frac{1 + T - 2.5 \times b(PD)}{1 - 1.5 \times b(PD)}, \tag{4}$$

where M is the effective maturity and T is the risk horizon of the investment project, $b(PD) = 0.00852 - 0.05489 * \ln(PD))$.

Parameters' shift and tilt for the effective period can be assessed on their own, for different types of investment projects on the basis of statistics. Also, the model can be adjusted to reflect the average duration of the investment project [8, 9].

Also, as a penalty, a factor of concentration of investment projects of the company can be considered, but the issue of simulation of concentration will remain outside the scope of this study.

4   Rating Model for Evaluation of Investment Projects

The general methodology for assessment of long-term sustainability of investment projects based on the calculation of the economic capital of the company is presented by the authors in several papers [3, 10, 11]. One component of this approach is to calculate the probability of default of investment projects analyzed by means of logit model.

Modeling the risks associated with investing is based on the equation of logit model. This method is widely used in theoretical research and practical prediction of defaults [10–13]. Logit model implies logistic transformation to the prediction data based on the maximum likelihood method [7].

In general, the logit model is represented by formula (5) [7]

$$PD = \frac{1}{1 + e^{-\left(b_j X_{ij}\right)}},\qquad(5)$$

where PD is the probability of default of the investment project, z is the linear combination of the factors of the regression model, X_{ij} is the value of the j indicator for the i investment project, and b_j is the regression parameter of the j factor.

The basis for logit models are data describing the financial performance of the investment project [4]. These include financial variables characterizing the model of the cash flows of the investment project, as well as a number of nonfinancial criteria to evaluate the type of project, its marketing component, and the experience of the company and project team in the implementation of similar investment earlier.

The result of using the logit model is the final ranking of investment projects according to their probability of default.

5 Evaluation of Economic Capital, *RAROC*, and Investment Decisions

Consider the investment program of an oil company with five investment projects. The initial parameters of these projects are presented in Table 1.

Cost management business risk involves determining the target level of financial stability, to maximize the value for a given level of risk. This level of financial stability can be defined as target long-term credit rating, which the company expects to receive. An important factor in the evaluation and management of risk becomes a value of the company and its development strategy. Each credit rating can deliver a certain level of probability of default, depending on the time horizon. One option corresponds to the rating, and the probability of default is presented in Table 2 [8, 9].

Probability of default determines the confidence level necessary for calculating the amount of unexpected losses and economic capital of the oil and gas company, which is calculated by formula (6)

Table 1 Main parameters of the investment projects

№	Projects	The total cost, mil. $	The project implementation period, years	Default probability, %
1	Overhaul of the pumping station	50	0.7	4.5
2	Oil field development	150	2	1.2
3	Construction of an oil warehouse	35	2	8.5
4	Modernization of refining at the oil refinery plant №1	120	4	5.5
5	Reconstruction of petrol station	30	2	5.4

Table 2 The correspondence between the probability of default and credit rating

Rating	1-Y PD	3-Y PD	5-Y PD
AAA	0.008%	0.03%	0.1%
AA	0.04%	0.16%	0.28%
A	0.16%	0.4%	0.58%
BBB	0.3%	1.4%	3%
BB	1.15%	8.6%	15%
B	5.8%	15.4%	32.6%
CCC or lower	26.57%	45.5%	60%

Table 3 Estimates of LGD for the main types of investment projects

Duration of the project/project type	Overhaul	Modernization of equipment	New construction
Short-time	12%	45%	65%
Long-time	30%	58%	80%

Table 4 Assessment of *RAROC* portfolio of investment projects

Projects	EL	ECAP	NI	RAROC	HR	RAROCspr
Overhaul of the pumping station	0.27	0.66	0.34	11%	10%	1%
Oil field development	1.44	3.40	1.92	14%	10%	4%
Construction of an oil warehouse	1.93	8.46	2.10	2%	10%	−8%
Modernization of refining at the oil refinery plant №1	3.83	25.82	6.93	12%	10%	2%
Reconstruction of petrol station	0.94	5.95	1.06	2%	10%	−8%

$$\gamma = 1 - PD \qquad (6)$$

where γ is the confidence, which determines the likelihood of non-bankruptcy, and *PD* is the level of probability of default corresponding to a target credit rating.

On the basis of distributions received, the LGD estimate is statistically different from basic parameters for each type of investment projects. The distribution is shown in Table 3.

Now, let us evaluate the *RAROC* for the purposes of making decisions about the implementation of investment projects. Assume the target level of financial stability is defined as rated BBB. The calculation of *RAROC* for each project is presented in Table 4.

Allocation of *RAROC* values for each investment project is used to determine which projects create business value and which destroy it. Thus, it is necessary to implement projects for overhaul of the pumping station, oil field development, and modernization of refining plant No. 1. The project on construction of an oil warehouse and reconstruction of petrol station is destroying the value of the business and its implementation of high risk and inappropriate. Of course, a number of projects are complex and embedded in the production processes; in such situations *RAROC* assessment can be made on a portfolio of investment projects, and the decision will be made regarding the multiple portfolios of investment projects.

6 Conclusion

In today's economy, characterized by high risks, maximizing the business value is inextricably linked to risk management. This problem is of particular importance for the oil and gas business, given the significant economic, technological and market risks.

Of particular importance, the model for investment management of business value proves its efficiency and ease of use, but a number of trends are forward-looking and provide opportunities for further development of the model. In particular, it is necessary to develop an approach to the evaluation of the correlation of investment projects with the general state of the economy, which involves the construction of multifactor indicator allowing us to identify global trends and their impact on investment activity. Another important issue is to improve the correctness of the model EAD, taking into account the distribution of the cost of the project at the time to default. One promising avenue is to develop a model *RAROC* segmentation in terms of impact on the value of the business and allocation *RAROC* on business units and divisions of the lower level. Solving these problems will allow for the maximization of the value of business under conditions of uncertainty and risk.

Acknowledgment The work was supported by Act 211 Government of the Russian Federation, contract № 02.A03.21.0006 and Russian Foundation for Basic Research (RFBR), contract № 16-06-00317.

References

1. Schroeck, G., Windfuh, M.: Calculation of Risk-Adjusted Performance Measures in Credit Markets, pp. 139–151. Schuling Verlag, Berlin (1999)
2. Schroeck, G.: Risk Management and Value Creation in Financial Institutions. Wiley, London. (2002)
3. Domnikov, A., Khomenko, P., Chebotareva, G.: A risk-oriented approach to capital management at a power generation company in Russia. WIT Trans. Ecol. Environ. **1**, 13–24 (2014)
4. Vasicek, O.: Loan portfolio value. Credit Portfolio Models. **15**, 160–162 (2002)
5. Ohlson, J.A.: Financial ratios and the probabilistic prediction of bankruptcy. J. Account. Res. **18** (1), 109–131 (2012)
6. Gorby, M.B.: A risk-factor model foundation for rating-based bank capital rules. J. Financ. Intermed. **25**, 199–232 (2003)
7. Merton, R.C.: On the pricing of corporate debt: the risk structure of interest rates. J. Financ. **29**, 449–470 (1974)
8. Gurtler, M., Heithecker, D.: Multi-Period defaults and maturity effects on economic capital in a ratings-based default-mode model. Finanz Wirtschaft. **5**, 123–134 (2005)
9. Gmurman, V.: Probability theory and mathematical statistics. High. Sch., 126–157 (2003)
10. Domnikov, A., Chebotareva, G., Khodorovsky, M.: Systematic approach to diagnosis lending risks in project finance. Audit Finance Analyses. **2**, 114–119 (2013)
11. Domnikov, A., Khodorovsky, M., Khomenko, P.: Optimization of finances into regional energy. Economy Region. **2**, 248–253 (2014)
12. Khodorovsky, M., Domnikov, A., Khomenko, P.: Optimization of financing investments in a power-generation company. WIT Trans. Ecol. Environ. **1**, 45–54 (2014)
13. Bellovary, J., Giacomino, D., Akers, M.: A review of bankruptcy prediction studies: 1930 to present. J. Financ. Educ. **33**(Winter), 31–43 (2007)

Rating Approach to Assess the Level of Investment Risks of Power-Generating Companies: The Case of Russia

A. Domnikov, G. Chebotareva, and M. Khodorovsky

1 Introduction

In modern conditions, characterized by a significant uncertainty, there is a need to develop analytical tools to solve the problem of increasing the accuracy of estimation of investment risks. Practical application of this tool will provide the investor with the necessary information about the risks, including those that are difficult to evaluate. Increased accuracy in risk assessment will allow the investor to rank investment projects tailored to their specific industry. The proposed methodological approach to improve the objectivity of the evaluation of investment risks has a direct impact on the structure of the investment portfolio and its profitability.

2 The Modern State System of Risk Management in Energy

A review of risk management systems in the historical context has shown that risk management-related issues became crucial and widespread in the middle of the twentieth century. That period of time saw the first publications devoted to comprehensive studying of risks and issues related to risk assessment and management. Among the authors who developed the basis of the contemporary risk management system are Markowitz [1], Sharpe [2], Smith [3], Merton [4], Gorby [5], Vasicek [6], etc.

In today's international practice, risk management is regulated by such basic international acts as the integrated risk management framework adopted by the Committee of Sponsoring Organizations of the Treadway Commission (the

A. Domnikov · G. Chebotareva (✉) · M. Khodorovsky
Academic Department of Banking and Investment Management, Ural Federal University,
Yekaterinburg, Russia

© Springer International Publishing AG, part of Springer Nature 2018
S. Syngellakis, C. Brebbia (eds.), *Challenges and Solutions in the Russian Energy Sector*, Innovation and Discovery in Russian Science and Engineering,
https://doi.org/10.1007/978-3-319-75702-5_14

COSO-ERM model), the risk management standard of the Federation of European Risk Management Associations (FERMA, the RMS model), and the standards adopted by the Bank for International Settlements (Basel II) [7].

The management system industry, including investment risks in power generation companies in developing countries, is characterized, as a rule, by the lack of a unique system of risk management [8]. A study conducted by the analyst firm KPMG [9] revealed a list of the most popular methods for quantitative risk assessment: scenario analysis, the method of Value at Risk, stress testing, gross margin-at-risk, etc. Also a classical method of simulation in the assessment of project risks is considered the Monte Carlo method proposed by Hertz in 1964 [10].

A number of analyses [9, 11] showed which issues are most relevant in the control system industry-specific risks in the electricity sector in developing countries to companies in the industry:

1. The 83% of companies have a documented policy of reducing industry risks
2. The lack of specialized bodies that implement a comprehensive system of risk management
3. Orientation of energy companies in the management of industry risk on the financial result for the period and not joint or carrying value
4. Limited use of hedging as a risk management tool
5. Limited use in predicting professional models, markets, etc.

However, different stages of development are methodological problems associated with the high level of subjectivity of assessing the level of risk and, in general, the investment attractiveness of companies in the industry. The solution of this problem by increasing the level of objectivity of the assessment lies in the methodological development of the mathematical apparatus that allows you to minimize the importance of expert opinion.

3 Methodological Approach to the Ranking of Investment Risks of Power-Generating Companies

Inputs to the study were reviewed by two groups of investment risks: exogenous and endogenous. The use of the above risk classification is due to the fact that the process of operation to the power-generating companies is influenced by external and internal factors. Differences in the level of their exposure and the degree of controllability by the company, in turn, cause a differentiated approach to the programming process for managing risks.

Tables 1 and 2 show how the group evaluates the risk and the performance of its calculation.

These exogenous and endogenous risks were selected for study from a larger volume of samples in accordance with the expert opinion on the basis of the survey

Table 1 Investment risks of power-generating companies: exogenous risks

№	Exogenous risk	The index for the calculation of risk
1	The risk of a lack of technological diversification	The index of technological diversification
2	The risk of insecurity in the region of the secondary energy resources	The share of energy generated by own sources of the region
3	The risk of energy efficiency	The share of decentralized systems in total energy consumption

Table 2 Investment risks of power-generating companies: endogenous risks

№	Endogenous risks	The index for the calculation of risk
1	The risk of increasing direct financial losses	Accounts receivable
2	The risk of dependence on imported equipment	The share of foreign equipment in the total amount of technical complex
3	The risk of depreciation of fixed assets	The share of obsolete fixed assets

of the general and financial directors of head and subsidiaries of Russian energy companies, as well as managers of risk management.

Exogenous risks are characterized by the following features [12, 13]:

- *The risk of lack of technological diversification* provides for a power-generating company in the irrational structure of the fuel balance. This can cause a possible reduction in quality of services and increases in production costs.
- *The risk of insecurity in the region secondary resources* involves the study of the degree of dependency of the region on resources coming from outside and, therefore, the negative effects caused by lack of own sources.
- The increase in the share of decentralized (autonomous) systems of energy generation reduces the competitive advantages of energy-generating companies and exacerbates the *risk of reducing the efficiency* of operation of such companies.

Analysis of endogenous risks revealed the following characteristics [12, 13]:

- The increased *risk of direct financial loss* is exacerbated in the situation of uncertainty of financial conditions of activity of a power company. In the field of energy generation it is the risk of insufficient work experience in a competitive market, difficulties in the implementation of investment programs, and low payment discipline of consumers.
- Modern technical equipment power-generating companies show a significant increase in the *risk of reliance on imported equipment*. On the one hand, the consequence of this situation is the increased expenditure on maintenance of equipment, receipt of consulting services in connection with the change of currency rates, etc. On the other hand, no domestic analogues make it impossible to implement a comprehensive program to mitigate this risk.

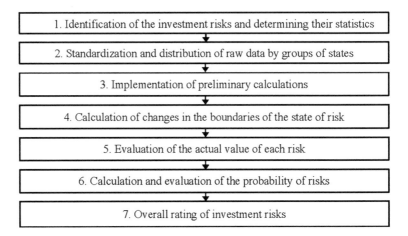

Fig. 1 Scheme methods of ranking investment risks of power-generating companies

- High yields of BPA and significant rates of obsolescence indicate a significant increase in the relevance of the risk of wear of BPA power-generating company.

The general scheme of the methodological approach to the ranking of investment risks is presented in Fig. 1.

The basis of this approach is the hypothesis that any risk assessment, basically, should be based mainly on the use of objective assessment tools, different from the opinions of experts.

The use of such mathematical instruments, as the Bayesian method, formulae, normalization, and statistical data as a source of information in the process of evaluating investment risks, allowed the authors to increase the independence of the final results of the study.

The necessity of normalization of the source data at the second stage of the method is associated with the existence of "distances in multidimensional space" and is aimed at bringing to comparability the studied indicators of investment risk.

In this study, the standardization of statistical data varies depending on the form of the impact of risk indicators on economic processes: indicators for direct and inverse proportion [13].

Valuation indicators in direct proportion are determined by the Eq. (1).

$$X_j^s = \frac{x_j - x_{\min}}{x_{\max} - x_{\min}} \tag{1}$$

where x_j is the actual value of the statistic, $x_{\min}$ is the minimum value of the statistic during the analyzed time series, and $x_{\max}$ is the maximum value of the statistic during the analyzed time series.

The third stage of the method associated with prior calculations involves the measurement values of the basic constituent elements, which are used in the calculation of the boundaries of the state change of risk, namely:

- Mathematical expectation (M_i) for each risk in each state
- Covariance matrix (S_i)
- The a priori probability of the objects of the class (q_i)
- Price erroneous attribution of objects to the class (c_i)

The calculation of the boundaries of the state change of each investment at risk in the general case is carried out according to Eq. (2) based on the Bayesian method [12]. According to this method for a set of objects subject to normal distribution, the object with the parameters X should be attributed to the aggregate of the first condition, if

$$
\ln\left(c_i q_i\right) - 0.5 \cdot \left((X - M_i)^T \cdot S_i^{-1} \cdot (X - M_i) - \ln|S_i|\right) - \left(\ln\left(c_{i+1} q_{i+1}\right) - 0.5 \cdot \left((X - M_{i+1})^T \cdot S_{i+1}^{-1} \cdot (X - M_{i+1}) - \ln|S_{i+1}|\right)\right) = 0 \tag{2}
$$

where X is vector of variables investigated in the space risks; M_i, M_{i+1} is mathematical expectation; S_i, S_{i+1} is covariance matrix, q_i, q_{i+1} are probabilities of occurrence of objects, and c_i, c_{i+1} is price erroneous classification of objects.

Assessment of the current status of each risk is carried out by comparing its actual value to a specific group status in accordance with the calculated threshold values. The level of each identified risk has a major impact in the overall rating of the investment risks. Preliminary ranking of investment risk at this stage of the methodology is primarily associated with the distribution of risk group by the assigned states.

The present study deals with four groups of state investment risk: minimal, acceptable, high and catastrophic level of impact risk. A brief description of these states is considered by the authors in several works.

The next step is the calculation of the relative index of each risk according to Eq. (3):

$$
Y_j = \frac{x_j^s}{x_j^{up}}, \tag{3}
$$

where Y_j is a relative indicator for each risk, x_j^s is the actual normalized value of each risk indicator, and x_j^{up} is the upper boundary of influence normalized value of the group risk.

The values of probability of investment risks are used as an additional factor in the ranking of risks in the case where the initial criteria coincide. The basis for calculation of the indicator is the statistical study of the dynamics of the metric data. The proposed method provides for the calculation of the maximum and the minimum of probability of risks.

The maximum likelihood estimate of risk is carried out according to Eq. (4):

$$
P_j^{max} = \frac{e_j^{max}}{E_j} \tag{4}
$$

where P_j^{max} is the maximum probability of risk $P_j^{max} \in [0; 1]$, e_j^{max} is the actual maximum number of adverse changes in the index of risk during the study period, and E_j is the total amount of change indicator of risk for the period.

The calculation of minimum risk base is not the full amount of adverse events, but only part of it is determined in accordance with the adopted horizon cutoff risk. The indicator is calculated using Eq. (5):

$$P_j^{min} = \frac{e_j^{min}}{E_j},\qquad(5)$$

where P_j^{min} is the minimum probability of risk $P_j^{min} \in [0; 1]$ and e_j^{min} is the adverse changes in the actual amount of risk for the period determined by the accepted cutoff horizon risk (G_s).

Thus, the presented methodological approach in the final rating of the investment risks according to their degree of risk for energy-generating company considers the following criteria in this order:

1. Belonging to the group level of influence of risk
2. The value of the relative measure of risk
3. The value of maximum probability of risk
4. The value of the minimum probability of risk

4 The Use of Rating Approach to Assess the Investment Risks of a Power Company

Consider the actual use case, which describes a methodological approach to the study of the branch of Russian power-generating company JSC "TGC-9," located in the Sverdlovsk region.

The source of information to estimate the statistics was the Ministry of Energy and Housing and Communal Services of Sverdlovsk region on the energy balance of the region for 2003–2014, as well as technological and financial statements of JSC "TGC-9" for 2005–2014. Indicators of endogenous risk for 2003–2004 were calculated based on the historical simulation method due to the fact that JSC "TGC-9" has started its activity since January 01, 2005.

The mentioned thesis defines one of the basic principles of the methodological approach implementation that is used for the distribution of values of indicators of investment risk for the four groups of States: the increasing level of risk from the increasing importance of defining the indicator.

The results of the calculations showed the following:

1. The first place on the level of danger to the power-generating companies shares the risks that constitute a group of a catastrophic level of risk: energy efficiency increases the risk of direct financial losses, the risk of dependence on imported equipment, and the risk of depreciation of fixed assets.

2. The fifth place is the risk belonging to the group of acceptable level of risk: the lack of technological diversification.
3. The minimum level of influence demonstrated risk of insecurity in the region's secondary energy resources.

5 Conclusion

As shown by the results of the study, methodological tools helped to solve the problem of subjectivity when making investment decisions. A distinctive feature of the ranking of investment projects is the application of Bayesian method. The final results showed that the highest risks faced by a potential investor will characterize the efficiency and the dependence on imported equipment. This, in turn, confirms the hypothesis of the advantage of risk assessment using objective assessment tools, different from the opinions of experts.

Acknowledgment The work was supported by Act 211 Government of the Russian Federation, contract № 02.A03.21.0006.

References

1. Markowitz, H.M.: Portfolio Selection: Efficient Diversification of Investment. Wiley, New York (1959)
2. Sharpe, W.: Portfolio Theory and Capital Markets. Economics, Boston (1970)
3. Smith, V.: Investment and Production. Harvard University Press, Cambridge, MA (1961)
4. Merton, R.C.: On the pricing of corporate debt: the risk structure of interest rates. J. Financ. **29**, 449–470 (1974)
5. Gorby, M.B.: A risk-factor model foundation for rating-based bank capital rules. J. Financ. Intermed. **25**, 199–232 (2003)
6. Vasicek, O.: Loan portfolio value. Credit Portfolio Models. **15**, 160–162 (2002)
7. Basel Committee on Banking Supervision. Proposed enhancements to the Basel II framework, www.bis.org
8. Domnikov, A., Chebotareva, G., Khodorovsky, M.: Development of risk management for power generating companies in developing countries. WIT Trans. Ecol. Environ. **193**, 859–870 (2015)
9. Market risk management in Russian electricity companies. Analytical study; KPMG (2012)
10. Hertz, D.: Risk analysis in capital investments. Harv. Bus. Rev. **41**, 95–106 (1964)
11. Domnikov, A., Khomenko, P., Chebotareva, G.: A risk-oriented approach to capital management at a power generation company in Russia. WIT Trans. Ecol. Environ. **186**, 13–24 (2014)
12. Domnikov, A., Chebotareva, G., Khodorovsky, M.: Evaluation of investor attractiveness of power-generating companies: special reference to the development risks of the electric power industry. WIT Trans. Ecol. Environ. **190**, 199–210 (2014)
13. Domnikov, A., Chebotareva, G., Khodorovsky, M.: Evaluation of investor attractiveness of power-generating companies, given the specificity of the development risks of electric power industry [in Russian]. Vestnik UrFU. **3**, 15–25 (2013)

The Application of International Risk Requirements for the Assessment of Investor Attractiveness of Russian Power-Generating Companies

A. Domnikov and G. Chebotareva

1 Introduction

The relevance of the study is due to the theoretical and practical significance of the international risk requirements to the evaluation of investment projects in the energy sector. The special role in modern conditions of economic growth and technological hazards requires the development of analytical tools and forecasting techniques of investment risks with the use of a system study tool. Furthermore, it contributes to the stimulation of growth of evaluation efficiency for power-generating companies' investor attractiveness.

Conducted by the authors, many years of research in this area showed that the assessment of emerging risks should accompany the development of specific mathematical apparatus that takes into account not only current requirements for risk management but also the features of the sector, and this is one of the main objectives of the present paper.

The result of the study is the authors' methodical approach in assessing the competitiveness of energy companies, which allows one to quickly identify industry risks and to assess their level of risk through the assessment of the value of these threats. The obtained results are of practical importance and are used in developing the strategy of development of power companies.

A. Domnikov · G. Chebotareva (✉)
Academic Department of Banking and Investment Management, Ural Federal University, Ekaterinburg, Russia

© Springer International Publishing AG, part of Springer Nature 2018
S. Syngellakis, C. Brebbia (eds.), *Challenges and Solutions in the Russian Energy Sector*, Innovation and Discovery in Russian Science and Engineering,
https://doi.org/10.1007/978-3-319-75702-5_15

2 International Risk Management Requirements

The desire to create uniform standards in the sphere of risk management began the process of developing international requirements to the system of risk management and implementing it worldwide. Prepared by the Basel Committee on Banking Supervision (BCBS), recommendations on risk assessment and management demonstrate the current trends in this sphere. Moreover, there is the basis for improvement of the risk management mechanisms, both credit institutions and the analogous departments of large industrial holdings.

Recommendations to risk assessment are presented in the framework agreement "International Convergence of Capital Measurement and Capital Standards" (Basel I 1988) and modified in the document Basel II (2004) and Basel III (2010) [1].

The main approaches to assessing investment risks BCBS proposes to introduce are [2]:

1. Standardized approach
2. The basic approach based on internal ratings (FIRB)
3. The advanced approach based on its own valuation models (AIRB)

The standardized approach is based on external credit ratings assigned by international rating agencies.

An improved version of the risk assessment is presented in basic and advanced IRB approaches where it is used for internal rating of the borrowers. And in the second case, BCBS offers to risk managers to measure the assessment components independently.

In turn, the need for the development and verification of the model under the AIRB stimulates the process of continuous development and improvement in this direction.

2.1 The Basic Components of a Risk Assessment in Accordance with International Requirements

International requirements proposed by the BCBS to the risk management system are associated with the use of basic risk components [1–4]:

- PD – probability of default. It characterizes the level of company risks in the implementation of the investment project.
- LGD – loss given default. It is the expected average cost of the company losses in the situation of default of the investment project.
- EAD – exposure at default. It characterizes the absolute value of the investment project sum and is determined by its full actual or forecasted cost of the investment, operating, and other expenses.
- M – maturity. This is the average duration of the investment phase of the project.

- ML_i – maximum losses of the i-th risk type.
- EL_i – the mathematical expectation of the i-th risk type losses.
- UL_i – unexpected losses of the i-th risk type.

3 Modified Approach to the Estimation of Investor Attractiveness of the Power-Generating Companies in the Implementation of the Investment Project

Proposed in the framework of the research approach to the assessment of investor attractiveness of a power-generating company is based on the well-known risk management theory of the economic capital and is the result of modification of a known Merton-Vasicek method [3, 4].

The economic capital is the amount of funds needed by the enterprise to cover the risks it faces in trying to maintain a certain standard solvency or in the event of default. In other words, economic capital allows maintaining by power-generating companies of the current level of their independence and stability and protection against economic losses as a result of implementation risks. Thus, economic capital determines the cost the company needs to have to provide a specified level of investment attractiveness and long-term sustainability.

A phased process of investor attractiveness assessment is presented in Fig. 1.

Initially, the study of the company investor attractiveness is preceded by identifying industry risks arising in the course of the project and describing its parameters. Source statistical data for each of the risk indicators are the basis for some of the parameters of risk calculation.

3.1 The Estimation Features of Risk Parameters Under the Modified Approach

The basic risk components used in the valuation include:

1. Average probability of default (PD)

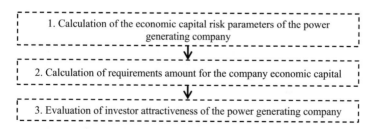

Fig. 1 Scheme of evaluation of the power-generating company investor attractiveness

2. Exposure at default (EAD)
3. Average loss given default (LGD)
4. Maturity (M)
5. Company's confidence level (α)
6. The level of correlation of company's condition with the region economy (r)

The average annual probability of default takes into account the totality of all financial and nonfinancial, including exogenous factors that influence to the project. In this study the probability of default is based on the standard form of a weighted arithmetic average given by the Eq. (1)

$$PD = \sum_{j=1}^{n} \left(P_j^m \cdot \gamma_j^s \right), \tag{1}$$

where PD is the investment project probability of default, P_j^m is the average probability of j-th risk realization, and γ_j^s is the j-th risk significance.

Exposure at risk is calculated as the full actual (or forecast) value of the investment project with the rate of return minus the value of highly liquid collaterals.

Assessment of LGD is based on analysis of statistical data for similar investment projects published in the area of default [5].

The maturity characterizes the effective period during which a stored position on the risk is determined by the duration of the investment phase of the project [4]. The maturity is evaluated using Eq. (2)

$$M = \frac{1 + T - 2.5 \cdot b(PD)}{1 - 1.5 \cdot b(PD)}, \tag{2}$$

where M is the maturity, T is the duration of the project investment phase, and parameter $b(PD) = (0.00852 - 0.05489 \cdot \ln (PD))^2$ is the investment project probability of default.

The company's confidence level is determined based on the assigned credit rating. The concordance between the expected probability of default (years) and rating is developed by international rating agencies and banking groups based on the statistics data.

The level of correlation of company's condition with the region economy status in this study is calculated based on the Pearson correlation coefficient given by Eq. (3) [6]

$$r = \frac{\mathrm{cov}\left(P_{jex}^{av}; P_{jen}^{av} \right)}{\sigma P_{jex}^{av} \cdot \sigma P_{jen}^{av}}, \tag{3}$$

where r is the correlation coefficient, $\mathrm{cov}\left(P_{jex}^{av}; P_{jen}^{av} \right)$ is the covariance value of the variables P_{jex}^{av} and P_{jen}^{av}, σP_{jex}^{av} is the standard deviation of the variable P_{jex}^{av}, σP_{jen}^{av} is the standard deviation of the variable P_{jen}^{av}, P_{jex}^{av} is the average probability of j-th exogenous risk realization, and P_{jen}^{av} is the average probability of j-th endogenous risk realization.

3.2 The Economic Capital Requirements for the Power-Generating Company

Under the modified approach to the estimation of the power-generating company investor attractiveness, the calculation of the initial values of the requirements for economic capital is carried out according to Eq. (4) [1, 7]

$$CaR = EAD \cdot LGD \cdot \left| N \cdot \left(\frac{N^{-1} \cdot (1 - PD) + N^{-1} \cdot (1 - \alpha) \cdot \sqrt{r}}{\sqrt{1 - r}} \right) - PD, \right| \quad (4)$$

where CaR is the requirements for power-generating company economic capital, N is the standard normal distribution, and N^{-1} is the inverse of the standard normal distribution.

In the case of exceeding the project investment phase duration by more than 1 year, there is a need for adjustment to CaR on the amount of the risk horizon as shown in Eq. (5):

$$CR = CaR \cdot M, \quad (5)$$

where CR is the requirements for economic capital subject to the penalty for the duration of the project investment phase.

4 Application of the Modified Approach to the Estimation of the Power-Generating Company Investor Attractiveness

Consider the actual case of use of the modified approach described on the example of Russian power-generating company JSC "TGC-9" in the implementation of the investment project "Construction of thermal power plant."

4.1 Scenario Approach to the Estimation of Investor Attractiveness

The framework of the studies considered two possible scenarios of development of the power-generating company investor attractiveness: "optimistic" and "pessimistic" [8].

Table 1 Scenarios for the development of the power-generating company investor attractiveness

№	Parameter's name	Scenarios	
		"Optimistic"	"Pessimistic"
1	Value of the investment project, bln rub	5	8
2	Investment project duration, years	2	3
3	Investor's income, billion rubles	2.2	3
4	Value of highly liquid collaterals, bln rub	2	1
5	LGD value, %	10	12
6	Cutoff risk horizon (G_s), %	1	50
7	The level of correlation (r)	0.5587	0.6879
8	Confidence level (credit ranking) (α)	0.9995 (A)	0.9964 (BBB)

Table 2 Probability of industry risk realization of JSC "TGC-9"

№	Risks names	Probability of risk realization				
		Maximum	Minimum		Average	
			$G_1 = 1\%$	$G_2 = 50\%$	$G_1 = 1\%$	$G_2 = 50\%$
Exogenous risks						
1	Risk of a slowdown in the development of the region industry	0.6364	0.1225	0.3130	0.3795	0.4747
2	Risk of reduction of the region investor attractiveness	0.2727	0.0125	0.1478	0.1426	0.2103
3	Risk of currency exchange	0.5455	0.1478	0.3215	0.3467	0.4335
Endogenous risks						
4	Risk of increase of the direct financial losses	0.8182	0.3625	0.5971	0.5904	0.7077
5	Risk of dependence on imported equipment	0.9091	0.2529	0.6799	0.5810	0.7945
6	Risk of wear of BPA	0.6364	0.1419	0.3418	0.3892	0.4891

The features of each of them are presented in Table 1: a direct impact on the assessment of the company investor attractiveness. Moreover, it plays an important role in the development of future scenarios for the company's investment policy.

4.2 JSC "TGC-9" Industry Risks in the Implementation of the Investment Project

As an example of assessment, the following exogenous and endogenous risks of the power-generating company considered are presented (Table 2) [8].

These are the results of the evaluation of the realization probability of exogenous and endogenous industry risks in the example of Russian power-generating company JSC "TGC-9."

The calculation of the maximum and minimum probabilities for each industry risk is based on the study of the dynamics of statistical data on their performance using the method of historical simulation. Estimation of the minimum probability also involves the use of scenario analysis when it changes the value of the cutoff risk horizon.

The results are shown in Table 2.

4.3 Evaluation of JSC "TGC-9" Investor Attractiveness

The final evaluation of investor attractiveness of the power-generating company considered is preceded by a calculation of risk parameters, as well as the requirements for economic capital.

Table 3 shows the values of risk components and requirements for economic capital of the company for each of the scenarios. The source data for their calculation are presented in Tables 1 and 2.

The results of calculations (Table 3) showed that the final requirements for economic capital of JSC "TGC-9" differ significantly from the proposed scenarios. So the requirements of the "pessimistic" scenario are higher than the analogous parameter by 4.4 times.

The main reason for the resulting differentiation is a "pessimistic" scenario's significant deterioration in terms of the economy functioning as a whole and the efficiency of the industry. This, ultimately, influenced the rise in the cost of the project, increasing the duration of its implementation, the requirements of investors to risk, etc.

The final decision about the current state of investor attractiveness is based on a comparison made to the company's requirements to cover risk (economic capital)

Table 3 Values of indicators of the investor attractiveness evaluation in accordance with the scenarios

№	Parameter's name	Scenarios	
		"Optimistic"	"Pessimistic"
1	Probability of default	0.4619	0.5881
2	EAD, bln rub	5.2	10
3	LGD, %	10	12
4	Maturity	2.0014	3.0091
5	Confidence level (credit ranking) (α)	0.9995	0.9964
6	The level of correlation (r)	0.5587	0.6879
7	Requirements for economic capital, bln rub	**0.4807**	**2.1236**

and the actual amount of funds that can be used in the event of a default of the investment project.

Study data of the financial statements of JSC "TGC-9" [9] showed that the actual amount of economic capital is 1.8 billion rubles.

Therefore, at this stage of research, it is difficult to draw a definitive conclusion regarding the attractiveness of a power company to investors. This is due to the fact that "optimistic" scenarios for the power-generating company are absolutely investor attractive: the actual amount of the funds exceeds requirements with a factor of safety of 1.3 billion roubles. This, in turn, can be used to cover other latent industry risks.

However, in case of execution of a "pessimistic" scenario, the company is recognized as investor unattractive: its own funds will not be sufficient to redress the full default of the investment project.

Thus, for the purpose of increasing the investor attractiveness and receipt of funds for the project of power-generating company, developing an appropriate program is required. Its activities usually involve two main areas [10–12]. In the first case, the acquisition of the required level of investor attractiveness is achieved through the attraction of additional funds to the amount not less than the deficit of the coating (in the studied case, 0.32 billion rubles).

5 Conclusion

The competitiveness of power companies in the investment market is largely dependent on compliance with international requirements for the risk management system. The special importance of these requirements is in their aim to implement capital-intensive projects that will improve the efficiency and reliability of power-generating companies.

The proposed modified approach to the estimation of investor attractiveness of power-generating companies based on the model of Merton–Vasicek has shown its effectiveness. In particular, the final requirements for economic capital of JSC "TGC-9" are significantly differentiated depending on the scenarios. This, in turn, improves the accuracy of evaluation of its investor attractiveness.

However, in terms of improving the methodological apparatus, it should be noted the necessity of considering the correlation of investment projects with the level of development and the trends in the economy.

Acknowledgment The work was supported by Act 211 Government of the Russian Federation, contract № 02.A03.21.0006.

References

1. Basel ommittee on Banking Supervision. International regulatory framework for banks, www.bis.org
2. Basel Committee on Banking Supervision. Proposed enhancements to the Basel II framework, www.bis.org
3. Domnikov, A., Khomenko, P., Chebotareva, G.: A risk-oriented approach to capital management at a power generation company in Russia. WIT Trans. Ecol. Environ. **186**, 13–24 (2014)
4. Domnikov, A., Chebotareva, G., Khomenko, P., Khodorovsky, M.: Risk-oriented approach to long-term sustainability management for oil and gas companies in the course of implementation of investment projects. WIT Trans. Ecol. Environ. **192**, 275–284 (2015)
5. Peter, C.: Estimating loss given default – experiences from banking practise. Springerlink. **2**, 143–175 (2006)
6. Gmurman, V.: Probability theory and mathematical statistics [in Russian]. High. Sch. **1**, 479 p (1997)
7. Gorby, M.B.: A risk-factor model foundation for rating-based bank capital rules. J. Financ. Intermed. **25**, 199–232 (2003)
8. Domnikov, A., Chebotareva, G., Khodorovsky, M.: Evaluation of investor attractiveness of power-generating companies, given the specificity of the development risks of electric power industry [in Russian]. Vestnik UrFU. **3**, 15–25 (2013)
9. Financial statements of JSC "TGC-9" for 2005–2014
10. Merton, R.C.: On the pricing of corporate debt: the risk structure of interest rates. J. Financ. **29**, 449–470 (1974)
11. Vasicek, O.: Loan portfolio value. Credit Portfolio Models. **15**, 160–162 (2002)
12. Ohlson, J.A.: Financial ratios and the probabilistic prediction of bankruptcy. J. Account. Res. **18** (1), 109–131 (2012)

Evaluation of Energy-Related Projects in Remote Areas

A. Domnikov, M. Petrov, and L. Domnikova

1 Introduction

The power industry of remote development areas is a specific section of Russia's energy policy. There is a strategic link between energy sector development and the development of new areas. It was the energy that has been helping Russia to expand into new territories for a long time. The key motives were the search for more natural resources and geostrategic organization of its areas. In today's Russia, these processes continue amidst the aftermath of drastic changes in basic social relations and new global challenges. Searching for the national strategy has turned into an intense struggle, which involves both internal and external forces. In this context, the strategic long-term development scenarios, including those for energy development, are determined by the progress and outcome of this struggle [1].

2 The Concept of Methodical Approach to the Evaluation of Energy Projects

Power facilities are characterized by increased capital intensity, a relatively long cycle of creation and subsequent operation, and strong direct and inverse correlations with the nature of development of the serviced territories and power

A. Domnikov (✉) · L. Domnikova
Department of Banking and Investment Management, Ural Federal University, Yekaterinburg, Russia

M. Petrov
Institute of Economics, The Ural Branch of the Russian Academy of Sciences, Yekaterinburg, Russia

© Springer International Publishing AG, part of Springer Nature 2018   133
S. Syngellakis, C. Brebbia (eds.), *Challenges and Solutions in the Russian Energy Sector*, Innovation and Discovery in Russian Science and Engineering,
https://doi.org/10.1007/978-3-319-75702-5_16

consumers. Creation of power facilities in remote development areas, which are mainly located in the Russian Arctic, including the Far North, enhances the specificity of feasibility thereof and influences the design. In this case, the energy sector plays a dual role. Firstly, it serves as a universal infrastructure for development and operation in the new territories and makes it possible to come to these territories and waters with economic purposes. Secondly, those areas of new development that have power resources can become a site for creation of the energy sector as a branch of specialization for the production, transformation, and transmission of energy to the consuming regions.

It is important to look for such relatively cost-effective solutions for the support systems, which would enable to reasonably reduce the energy costs in an adequate, reliable, and safe way. This requirement is a major feature of the design and of application of the economic criteria used for substantiation of the development solutions in this area.

Whereas the energy burden on the economy of a developed territory with developed economic complexes is made of the total energy costs of value-added chains which creates a mechanism for management and limitation of such costs, the areas of new development have no such restrictions and no built-in mechanism of cost management and limitation.

Hence, on the one hand, feasibility of cheaper energy infrastructure within the specified constraints is quite evident. On the other hand, time lags of the infrastructure development can appear longer than expected. It requires the correct configuration of the mechanism for addressing the time factor. It shall be taken into account that during the life cycle of the energy system, the directions of its development may change due to the appearance of new objectives and infrastructure objects, which means increased uncertainty in development of new areas in the North and requires more reserves.

With that said, it can be stated that the economic criteria do not have the highest priority for the infrastructure of new territories in general, including their energy sector, which increases the role of multi-criteria approaches.

In our view, the following principles of criterial consistency in the hierarchy of the systems shall be reckoned as the most important for making decisions on development of the energy system of the remote areas [1]:

- The results of the use of resources invested in the development of the system exceed the bounds of the system (technical, economic, political, logistics).
- It is possible to evaluate the use of these resources only in the context of objectives external to the infrastructure systems.
- Resource constraints of the energy systems depend on the resource potential of the country and region.

Subject to the foregoing, the issues of comparative effectiveness of energy systems, when an economic decision-making criterion is manifested indirectly, have their independent value; this criterion is used to streamline the system at the decision-making stage when its production capacity has been already determined based on well-reasoned scope of needs. For example, the need for energy is set using the initial parameter that is dynamically projected – energy consumption and its structure in terms of energy resources, facilities, and location.

The issue of comparative effectiveness is addressed with the purpose of selecting and ranking project options that meet the restrictions and more preferred criteria. Methodologically, it is important that the comparative effectiveness is considered after forecasting the conditions when the effect of volume indicators defined in the course of time is equivalent to the structure of the consolidated energy needs which in this case can be considered independent of fluctuations in energy costs. Actually, under given conditions of the equivalence of this effect, the issue of comparative effectiveness of the use of resources in the infrastructure system is resolved. The idea is to find a rational trajectory and composition of resources to ensure certain conditions of the effect equivalence.

Development issues for large systems are normally long-term issues, when it is necessary to consider a long optimization period. This is mainly due to both the capital intensity of the systems under consideration and systematicity as such. The planning horizon reflects the aftereffect: forward costs are to be borne to meet the needs of subsequent phases of the system development.

Due to the longtermness, a special role in the behavior of the economic criterion belongs to the time factor, i.e., the need to take into account changes in the relative value of system resources in time. The time factor is taken into account through intertemporal reduction of results and costs.

The methodology for calculation of integrated UNIDO cash flows that is the basis of the entire common practice of investment appraisal, involves the same approach to one-time and ongoing costs and their discounting using the same factors [2]. For the tasks described here, this approach seems inadequate, because it reflects only the value of money over time which is not the decisive factor for the tasks of strategic nature. In this case, private investments may go to the new development regions as an alternative of their application in other regions. Outside of such situation which is still very new for funding new development in this country, universal discounting at the market rate of interest gives distorted indications since it is practically not linked with the value of the resource potential created in the result of the territorial development.

In the North, the importance of adequate development of the energy sector is particularly high due to the harsh climate, territorial remoteness, and increased energy intensity of production. However, for the newly developed territories in the North there is a greater variance in building energy schemes, since isolated solutions based on the use of local fuels are possible. Spot and ribbon development options increase the capabilities for the development of decentralized (distributed) power industry.

For areas of the North, the Nether-Polar, and the Polar Urals lying at the junction of the constituent entities of the Russian Federation, the trans-regional approach is needed which allows to directly compare the energy solutions based on the potential of both Cisuralian and Transuralian entities of the Russian Federation. At the same time, these solutions need to be incorporated in the regional energy strategies, programs, and development schemes. Elaboration of such decisions in the regions is becoming more and more in demand in the modern context.

The regional energy strategy should become the basis for organization of integrated regional power industry development management. This means that implementation of the energy strategy is possible subject to further elaboration and approval of regional and interregional target power industry development programs forming pipelines of interrelated and adequately resourced projects in a number of areas. Such areas include:

- Preparation of conditions for organization of large-scale energy construction with long-term prospects
- Development of cogeneration
- Development of network infrastructure
- Development of local and small-scale power generation
- Energy saving
- Optimization of the fuel and energy balance of the region

Availability of approved programs will enable maximum possible application of incentive mechanisms to encourage energy-generating and energy-consuming entities to invest in energy generation, technological progress, and energy control and ultimately to attract to the regional power industry development the resources of sectors and actors operating both in the region and in the neighboring territories bearing in mind interregional importance and effects of large-scale activities in the energy sector. On the basis of the strategies and programs, initiatives are to be selected, and the state support is to be provided to the projects in the area of energy supply of the region and its individual parts. Scientific engineering and forecasting analytical support of such programs should be based on feasibility reports on energy industry development and on deployment schemes regularly developed by specialized organizations. Based on such schemes systematic design and project development cycles for individual energy facilities shall be arranged. Thereby the contour of integrated management of energy sector development shall be restored and substantially upgraded.

Strategic energy management is based on the principles reflecting the operation and development of any regional system within the Interconnected Energy System (IES) of the macro region of the Urals, Siberia, Northwest, and the United Energy System of Russia. Therefore, the regional energy strategy takes as a premise the priority of the operation of the regional power grid in the interconnected and unified power grids. The system methodology applied to the development of the power industry implies that the substantiation for the development of power generation facilities and backbone electrical networks should assume effective operation of large electric power systems, i.e., the decisions on major energy facilities and system interconnections should be assessed relative to the reference system on the interregional level, in this case, Urals IES or Northwest IES. Absolute priority in managing the development of the power industry belongs to the national level. The unified power system has the highest systemic status in comparison with other infrastructures and possesses all attributes of a large system. Its reproduction as such is the basis of effective development of all business units and the basis of the national security. However, this does not exclude regional management of the power industry development.

Firstly, in the energy sector, there is a technological possibility and feasibility for using small capacity and local (stand-alone) systems. Secondly, the regionalization of the economy has led to concentration of significant development resources on the regional level. And, thirdly, these processes take place against the background of weakening of the country's power industry management as a result of economic reforms. The regions in these circumstances cannot be guaranteed adequate power supply status "from above."

In the North, the effective energy sector development is possible through a combination of capabilities of large- (systemic) and small-scale (mostly distributed) energy generation. The main prerequisites for this are the focal character of loads and significant demand for heat in the development areas. In both cases, cogeneration of electricity and heat makes economic sense.

The strategic development priorities for a distributed energy sector in the new development territories are no less important. Historically, this type of facilities performed the role of pioneer energy industry in the Far North and equivalent territories. In the areas with available natural and associated gas, the bases of the energy sector were gas turbine installations with unit capacity up to 16 MW, while in other regions were diesel generators using motor fuel. In the course of the development of new territories, the growing power loads were routinely connected to centralized sources, namely, the district energy systems.

In the areas currently the segment of distributed power generation is rapidly growing. It is based on energy sources, constructed and operated by consumers. Solutions are becoming more competitive in the situation of accumulated underdevelopment of the large power industry and the growth of the tariff burden on the economics of the consumers, on the one hand, and improvement of technological capabilities of small (distributed) power industry on the other hand. Practically, the most significant overlapping of the opportunities of the large and small power generation today appears to be a set of heat and electricity cogeneration technologies.

3   Performance Evaluation of Cogenerating Power Plants

The preliminary analysis of the business processes related to the performance evaluation of cogenerating power plants revealed that the following four options out of all abovementioned ones are the most promising in terms of their efficiency [3]:

1. Modernization of cogenerating power plants with the replacement of the worn-out elements, which are mainly used in the areas with high temperatures and pressures

2. Re-equipment of cogenerating power plants while preserving their existing dimensions
3. Expansion of the existing cogenerating power plants through installation of additional cogeneration CCGT units in the new main buildings
4. Construction of new cogenerating power plants based on CCGT units equipped with solid fuel gasifiers

In general, assessment of economic efficiency requires additional calculations with the use of modern mathematical tools of fuzzy set theory, which allows solving multi-criteria problems and eliminates the inconsistency between some indicators when choosing the best variant for development of a cogenerating power plant. This multi-criteriality may be considered as a manifestation of uncertainty connected with the conditions of development and future operation of a cogenerating power plant [4–6].

To solve this problem, we have developed a general method of multi-criteria analysis, which consists of the following steps [3, 5]:

• Identification of admissible alternatives among many options (at this stage we select the most preferable development alternatives, which satisfy the conditions of the problem being solved)
• Definition of a set of criteria (objectives), which may help to solve the problem
• Determination of the criteria for evaluation of feasible alternatives in exact numerical values, interval (fuzzy) numbers
• Compromise allocation (a set of compromises may be applied only to the options for which the best combination of all criteria cannot be found);
• Application of the fuzzy set theory for identification of the non-dominated set of alternatives, which may determine their effectiveness
• Analysis of the calculation results, adoption of decisions

Comparison of the options, which may be applied to cogenerating power plants, was performed on the basis of the integrated suite of software facilities developed by the authors and used for the fuzzy multiple criteria analysis [7–10].

The calculations were performed within a predetermined range of possible changes in the weight coefficients talking into account certain groups of criteria. The increment range was 0.25.

Figures 1, 2, and 3 show the results of the multi-criteria analysis, performed with the use of mathematical tools of fuzzy set theory, and hierarchical grouping of the alternatives by the degree of their non-domination depending on the weight coefficients.

With a relatively wide range of weight values in different groups of criteria (0.25–0.5), including the case of equal probability of these groups, the most effective was the option, which involved expansion of the cogenerating power plant through addition of a CCGT unit. Compared to other business processes, the same option has the greatest degree of non-domination in terms of energy saving and environmental protection (with the weights of these criteria equal to one). Under the same conditions, the second option is construction of a new cogenerating power plant on the

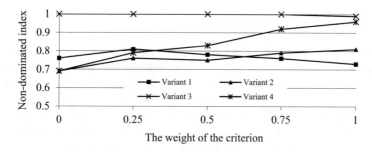

Fig. 1 Ranking of the alternatives by the energy criterion

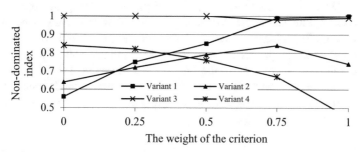

Fig. 2 Ranking the alternatives by the economic criterion

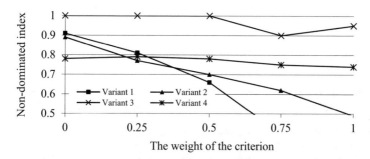

Fig. 3 Ranking the alternatives by the environmental criterion

basis of a CCGT unit equipped with a solid fuel gasifier. The analysis shows that the lower rank of this option may be explained by the following reasons: as for the energy criterion, by the significantly smaller volumes of power generation in the initial stage (due to longer construction period), and as for the environmental criterion, by the need in additional alienation of land. With the increase of the economic criterion weight (0.75–1), the equipment modernization becomes the most effective business process, which is also the best investment solution.

The result of multi-criteria analysis shows that the most effective are options 1 and 3. The choice between these alternatives may be made based on the existing strategic and tactical plans adopted by the management entities responsible for the development of the centralized cogeneration system.

4 Conclusion

The economic and technological development priorities in production of combined heat and power energy in remote areas were assessed with the use of mathematical tools of fuzzy set theory. The results of this analysis are as follows: (1) taking into account a short-term perspective, for the centralized cogeneration systems, the most preferable will be extension of the residual operation life of the main equipment; but in terms of a long-term perspective, it seems to be more effective to expand the existing plant capacities through construction of additional CCGT units, which will remain competitive under the conditions of seasonal fluctuations in the consumer heat load; (2) for distributed power generation systems, re-equipment of the existing boiler stations with additional installation of CCGT units proved to be the most competitive option. Electric capacity of such power plants will cover the annual needs in hot water supply for consumers, while the plants will be able to work in economy mode during the non-heating season. It was established that in the remote areas, which possess local fuel resources (waste wood, agricultural wastes, etc.), it is feasible to use high-performance gas-producing cogenerating power plants equipped with internal combustion engines.

Acknowledgment The work was supported by Act 211 Government of the Russian Federation, contract № 02.A03.21.0006 and Russian Foundation for Basic Research (RFBR), contract № 16-06-00317.

References

1. Tatarkin, A., Petrov, M., Litovskiy, V.: The Arctic Urals in the system of congruous areas: Scientific approaches and economic practice. Prob. Anal. Public Admin. **6**(38), Vol. 7, 25–44 (2014)
2. Berens, B., Havranek, P.: Guidelines for Assessing the Effectiveness of Investment, pp. 127–154. INFRA-M, Moscow (1995)
3. Domnikov, A.Y.: Aspects of multi-criteria analysis for technical re-equipment of power plants [in Russian]. Ser. Econ. Manag. Ekaterinburg Ural State Tech. Univ. **1**(53), 25–36 (2005)
4. Braid, R.B.: The importance of cumulative impact assessment and mitigation. Energy UK. **5**, 643–652 (1985)
5. Domnikov, A.Y.: The strategy of increasing the competitive advantages of the territorial system of cogeneration energy, [in Russian]. Sci. Inform. J. Econ. **1**(38), 35–47 (2008)
6. Domnikov, A., Chebotareva, G., Khodorovsky, M.: Systematic approach to diagnosis lending risks in project finance [in Russian]. Audit Finan. Anal. **2**, 114–119 (2013)
7. Ward Jr., J.H.: Hierarchical grouping to optimize an objective function. Am. Stat. Assoc. **58** (301), 236–2446 (1963)
8. Khodorovsky, M., Domnikov, A., Khomenko, P.: Optimization of financing investments in a power-generation company. WIT Trans. Ecol. Environ. **1**, 45–54 (2014)
9. Domnikov, A., Khomenko, P., Chebotareva, G.: A risk-oriented approach to capital management at a power generation company in Russia. WIT Trans. Ecol. Environ. **1**, 13–24 (2014)
10. Domnikov, A., Khodorovsky, M., Khomenko, P.: Optimization of finances into regional energy. Econ. Reg. **2**, 248–253 (2014)

Tax Sources of Funding the Road Network as a Tool to Increase Transport Energy Efficiency

I. Mayburov and Y. Leontyeva

1 Introduction

Over the past 20 years, Russia has had to deal with an explosive growth in the number of privately owned cars and an increasingly heavy load on roads. According to government data, the poor quality of roads leads to a loss of around 3% of GDP annually. Substandard roads are the primary reasons for this as poor road surface quality pushes fuel consumption up by 30% and increases the cost of road transport by 50% compared to those in European countries. This conclusion echoes the Global Competitiveness Report 2013–2014 by the World Economic Forum that placed Russia 136th (out of 144) in the world for the quality of roads (international experts scored road quality in Russia at 2.3 out of 7) (Schwab and Xavier [1]).

Long distances and underdeveloped roads increase the cost of domestically produced goods and make them less competitive. On average, transportation costs make up 15–20% of the ultimate cost of production in Russia, compared with 7–8% in developed countries.

Federal motorways constitute the framework of the national road network as they provide international and interregional links. The quality of federal motorways has always been higher than that of other types of roads (regional, local ones) but is still very poor. Twenty-five percent of federal motorways are overloaded, 4% have reached their traffic capacity limit, 56% of federal motorways have pavement strength issues, and 37% do not meet roughness standards. Only 8% of federal motorways have four and more lanes. Eight percent of federal roads are still surfaced

I. Mayburov (✉)
Ural Federal University, Russian Federation, Ekaterinburg, Russia

Far Eastern Federal University, Russian Federation, Vladivostok, Russia

Y. Leontyeva
Ural Federal University, Russian Federation, Ekaterinburg, Russia

© Springer International Publishing AG, part of Springer Nature 2018
S. Syngellakis, C. Brebbia (eds.), *Challenges and Solutions in the Russian Energy Sector*, Innovation and Discovery in Russian Science and Engineering,
https://doi.org/10.1007/978-3-319-75702-5_17

141

with gravel. As a result, 60% of federal motorways are substandard. In areas of the north, Siberia, and the Far East, a backbone road network connecting all regions of Russia is yet to be built.

When it comes to local and regional roadways, the situation is even more dismal. Some 46,000 settlements with a combined population of around 2.7 m have no paved road connection to the national road network.

It is quite clear that forced improvement in the quality of the national road network is needed to make the Russian economy more competitive. This strategic task requires a larger amount of investment than earmarked in the federal budget. Funding for investment has to be built up for a long-term period rather than as a lump sum. Against the background of an economic crisis and low natural gas and oil prices that Russia largely depends on for revenue, the government is short of resources for road construction. This makes the task of finding a more effective mechanism of accumulating and distributing funds for road construction even more timely and relevant. It is equally important to find new fiscal sources of revenue for road funds that would in the long-term perspective shift the financial burden of road development to car owners (Botlikova et al. [2]).

2   Analysis of Sources of Funding for a Road Network

A peculiar feature of public expenditure on road infrastructure over the past 15 years has been the fact that its dynamics are reflective of the country's economic development and are even slightly forward-looking. Growth in spending on the road network is in direct correlation with GDP and state revenue growth (Table 1).

Although there are positive dynamics of the road construction cost in Russia, the authors have to note the inadequacy of the cost. At present the share of the road construction cost has reached 1.8% of the Russian GDP. This indicator is significantly lower than that of other European countries. For example, in Italy, the share of the road construction cost has reached 4.8% of GDP, Finland 4.3%, the UK 4%, Spain 3.9%, France 3.7%, Sweden 3.2%, Norway 3.1%, Denmark 2.9%, and Austria 1.9% (Doll and Van Essen [3]).

The increase in the spending to 1.8% of GDP is insufficient because, unlike the well-established quality road networks in developed countries, Russia's road infrastructure is fraught with imbalances.

First, vast territories in the north, Siberia, and the Far East are void of any road construction projects and practically lack a federal road network that would provide uninterrupted connection between all regions. There is also a drastic shortage of paved regional and local roads that would be passable all year round.

Second, maintenance costs are higher because of higher wear and tear to the surface to overloaded roads. The costs are also high because of the damaging impact of extreme temperatures ranging from 40 to −40 °C to the road surface.

Third, the quality of the existing roads and infrastructure is very poor and does not meet international standards.

Table 1 Dynamics of revenues from transport taxes and spending on road infrastructure

Indicator, billion RUB	2000	2005	2010	2011	2012	2013
Gross domestic product	7306	21,620	46,309	55,800	62,599	66,755
1. Combined revenues of the Russian federal budget, including:	*2097.7*	*8579.6*	*16,031.9*	*20,855.4*	*23,435.1*	*24,442.7*
Fuel taxes	n/a	125.5	169.8	283.5	365.8	418.2
Car excise tax	n/a	5.1	15.1	26.1	33	33.3
Transport tax	n/a	26.0	75.6	83.2	90.2	106.1
Recycling fee	n/a	0	0	0	18.7	49.5
Total for transport-related taxes	n/a	156.6	260.5	392.8	507.7	607.1
Share of transport-related taxes in combined budget revenues, %	n/a	*1.83*	*1.62*	*1.88*	*2.17*	*2.48*
2. Total government spending	*1960.1*	*6820.6*	*17,616.7*	*19,994.6*	*23,174.7*	*25,290.9*
Spending on road infrastructure, including:	40.3	250.5	645	714.2	990.5	1172.3
Spending by federal government	1.9	42.1	281.1	349.5	442.4	504.5
Spending by regional governments	38.4	208.3	363.9	424.5	646.3	731.2
Share of spending on road infrastructure in total government expenditure, %	*1.92*	*2.92*	*4.02*	*3.42*	*4.23*	*4.80*
Share of spending on road infrastructure in GDP, %	*0.55*	*1.16*	*1.39*	*1.28*	*1.58*	*1.76*

Fourth, there is a lack of private investment in toll roads as business is not interested in making capital investment in road development because of a long payback period and a low return.

All these are factors that necessitate further growth in public spending on road infrastructure. One of the ways to ensure the growth is by enhancing the fiscal value of transport taxes in Russia.

3 Types of Transport Taxes and Charges in Russia

The following levies and charges make the framework of transport-related taxes in Russia (Table 2).

Car sales taxes are paid by the manufacturer (or the importer) on the sale of cars and motorcycles. The selling of lorries and buses is exempt from excise taxes. The tax rate is progressive.

Fuel taxes are paid by the producer and are included in the price. Tax rates vary for Class 4 and Class 5 fuels. Excise taxes on fuel are designed to establish a link

Table 2 Types of transport taxes and charges in Russia

№	Levy	Connection between amount paid and intensity of road use by car owner
1	Car sales taxes and fees	No
2	Fuel tax	Yes
3	Motor oil excise tax	Yes
4	Transport tax	No
5	Parking fees	No
6	Recycling fee	No
7	Heavy vehicle use tax (gross weights equal or exceeding 12 tones)	Yes
8	Mandatory and voluntary insurance	No

between the mileage of the car and its owner's financial contribution to road funding in the area of fuel purchase. This connection works in the following way: fuel tax revenues are accumulated in the federal treasury and distributed among regions for purposes of road construction. The funds are allocated in proportion to fuel consumption in each region. This arrangement implies a strong connection between fuel consumption in the region and traffic load on the road network there. The advantage of the tax is that it strongly correlates with traffic load on the road network. Excise taxes on motor oils are paid by the manufacturer or the importer and are included in the price (Magaril et al. [4]).

Transport tax is essentially an ownership tax. The transport tax rate is based on engine horsepower. Tax rates vary for cars, lorries, buses, motorcycles etc. and are highly progressive.

Parking fees are paid by car owners for being able to park their vehicle in a designated place (usually in downtown areas) during a specific period of time.

In Russia, the recycling fee is imposed on all vehicles and is paid before their registration with the traffic police. The fee guarantees that the owner will not have to pay for scrapping his car upon the end of its service life. The fee is paid by car manufacturers or importers.

The heavy vehicle use tax is levied on heavy freight vehicles with registered gross weight exceeding 12 tonnes for operating on public motorways. It is designed as compensation for damage that heavy vehicles cause to roads. The tax will be determined according to the length of the stated route that will be tracked by means of GPS navigation.

The analysis shows that of the eight types of fees imposed on car owners in Russia, only three appear to establish a direct connection between the amount paid and intensity of road use by the car owner.

The main difference between the system of transport taxes in Russia from that in developed countries is that car owners do not appear willing to increase their financial contribution to road network funding. The opportunistic attitudes among the majority of taxpayers are due to their conviction that their payments dissolve in combined budgets and are not specifically spent on the development of the road network. Funding for roads in a majority of countries is intrinsically linked to a system of taxes on road users through so-called road pricing. It defines the key principle of road user charges that are designed to encourage the reduction of users' environmental impact, the preservation of transport links, and higher efficiency of the road network, including by means of international shipping regulations. Experience has shown that vehicle levies and road charges prove most effective when the payment for road use is made in the closest proximity to the point of consumption and the size of the charge is set at a level that is close to marginal costs of road maintenance.

4 Establishing Links Between Transport Charges and the Development of Road Networks

Establishing specific links between transport-related charges and the development of road networks is a fairly complicated process. In international practice, the following variable cost charges, that is, those depending on the intensity of road network use, are typically applied (Delucchi [5]):

- Fuel taxes are excise taxes levied on fuel sales.
- Link tolls are a fee for using a particular link (motorway).
- Point tolls are charged for using a particular road facility (a bridge, tunnel, or ferry).
- Distance-based network charges are levied on any trips made on roads.

In Russia the most commonly used charge is fuel excise tax. It is the most acceptable fiscal instrument that makes tax payments proportional to the amount of road use.

Fuel excise taxes are only compensatory; they have no impact on car owners and cannot encourage or discourage them from using cars and opting for public transport instead. One tax impacting taxpayers' behavior is vehicle tax with a variable rate for urban rural areas. Its size increases progressively if the taxpayer buys a second or third car.

With the absence of factors discouraging taxpayers from owning a vehicle, the community will fall into an institutional trap that sends communities into a spiral of automobile dependency: car ownership costs are low, and an increasing number of community members use private cars for their transportation needs. The level of car ownership keeps rising, necessitating the construction of new roads. Public transport cannot compete with private cars and falls into decline; cities become unfit for non-motorized traffic and recreation, people start moving to the suburbs, and daily long-distance trips require more roads. A large number of cities in the USA have already found themselves in the trap, Detroit, Dallas, and San Jose among them. These cities are designed for car users only.

5 Avoiding Falling into the Institutional Trap of Car Dependency

To avoid falling into the institutional trap of car dependency, it is necessary to set variable prices of owning and running a car in territorial communities with different rates of car ownership. The tax price should be high in communities where the car ownership rate is high. For implementing the benefit principle, it is necessary to make transport-related fees purpose specific: they should be paid to motorway funds and spent on the development of road networks.

It is necessary to make sure that car owners pay for the negative externalities. Car owners should also reimburse government for funding the maintenance and development of road networks. They may disagree with the idea because their internal costs of buying, servicing, and driving the car are quite high. Converting the external costs that are now paid for by the entire community into internal costs that are borne by car owners will inevitably result in diminished interest in owning a car and stronger interest in using public transport. This adjustment of consumer behavior should be a goal for urban areas.

A considerable increase in transport tax rates could provide an escape from the trap. But the hike should be differentiated for different territorial communities depending on the rate of car ownership and road density, as well as environmental characteristics of motor transport. And the community itself should be involved in determining the tax rate. The rates should be different not only within the region, because the factors of car and road use might vary drastically in the regional capital and in rural areas. Territorial communities should also be provided with targeted sources of funding for road construction and maintenance (De Borger [6]).

6 Increasing the Fiscal and Regulatory Significance of Transport Taxes

Transport tax has to have a regulatory role because car ownership cannot grow unchecked. Transport tax must not be neutral to people's investment decisions when buying and owning a car. In one territory it should encourage car ownership, while in the other, on the contrary, the tax should discourage it and create alternative solutions including public transport. The regulatory function can only be fulfilled through a considerable differentiation of rates on the basis of territories as well as on means of transport.

We consider it expedient to reform transport tax without any radical change to the tax base made up of engine capacity since a state registry has already been created and tax administration processes have already been fine-tuned. We suggest that changes should be implemented with regard to tax rates, the nature of the tax, and how its elements are managed on different levels (Mayburov and Leontieva [7]).

First, transport tax should become a special purpose tax. Revenues from the tax should be accumulated in road funds of regions and municipalities. The use of funds should be restricted to the purposes of environmental protection, road construction, and maintenance.

Second, the tax base of transport tax should be jointly used by regions and municipalities. The tax rate should, therefore, be split into regional and local components.

Third, three adjustment factors could be applied to the regional and municipal components of the transport tax rate: vehicle environmental class, car ownership rate, and road density. The adjustment factors can increase as well as decrease the appropriate part of the tax rate.

Fourth, the car ownership factor should adjust transport tax rates with respect to the number of cars in use in a particular area (region or municipality). For example, if the car ownership rate is high (over 300 units per 1000 inhabitants), a higher adjustment factor of 1.5 should be used to discourage car ownership. If the car ownership level is medium (200–300 cars per 1000 inhabitants), the factor is 1. For areas with low car ownership, a reduction factor of 0.5 should be applied.

Fifthly, the road density factor should adjust transport tax rates, bringing them into line with the development of road infrastructure. Car owners in areas enjoying better roads should pay a higher tax rate, while those with a less developed road network should pay less. For example, if the road density is high (over 500 km of roads per 10,000 sq.m.), a higher adjustment factor of 1.5 should be applied. For areas with a low road density (less than 200 km per 10,000 sq.m.), there could be a reduction factor of 0.5.

Along with car regulation through taxation, various means of public transport development should be used. More specifically, public transport, including taxis, should be eligible for tax concessions or even exempt from some vehicle charges. Permission should be granted to use money from motorway funds for subsidizing public transport in order to make it more attractive to people pricewise.

7 Conclusion

Today, transport tax in Russia is fiscally insignificant. It has no influence on people's investment decisions as to whether to own a car. It does not take into account vehicles' environmental class and provides enough funds for road development in regions and municipalities. As a result, road construction in regions and municipalities is very slow.

In the paper we have proved that in order to escape from this institutional trap, transport tax has to be reformed so that its fiscal function enhances, and it acquires a regulatory role.

A hike in the tax price of using automobiles will restrict the growth of car ownership. Tax concessions and subsidies for public transport will spur the development of public transport and make it more affordable.

The poor quality of roads in Russia negatively impacts population mobility and drives up the cost of production, car crash, road traffic death, and injury rates. This leads to considerable economic losses that might amount to 10% of GDP. At the same time, spending on road infrastructure increases annually but never exceeds 2% of GDP. An increase in the development of the road network might lead to considerable economic benefits for the public purse in the future and improve the energy efficiency of transportation.

Acknowledgement The work was supported by the Humanitarian Scientific Foundation of the Russian Federation, contract № 17-22-21001.

References

1. Schwab, K., Xavier, S.M., The Global Competitiveness Report 2012–2013, World Economic Forum, The Global Benchmarking Network. Geneva (2012)
2. Botlikova, M., Botlik, J., Vaclavinkova, K.: Negative impacts of transport infrastructure funding. In: Creating Global Competitive Economies: 2020 Vision Planning and Implementation Proceedings of the 22nd International Business Information Management Association Conference in Italy, Rome, vols. 1–3, pp. 1141–1152 (2013)
3. Doll, C., Van Essen, H.: Road infrastructure cost and revenue in Europe. Produced Within the Study Internalization Measures and Policies for all External Cost of Transport (IMPACT), Deliverable 2, Delft (2008)
4. Magaril, E., Abrzhina, L., Belyaeva, M.: Environmental damage from the combustion of fuels: Challenges and methods of economic assessment. WIT Trans. Ecol. Environ. **190**(2), 1105–1115 (2014.) WIT Press: UK
5. Delucchi, M.: Do motor-vehicle users in the US pay their way? Transport. Res. Part A. **41**(10), 982–1003 (2007)
6. De Borger, B.: Optimal congestion taxes in a time allocation model. Transport. Res. Part B Methodol. **45**(1), 79–95 (2011)
7. Mayburov, I., Leontieva, Y.: Transport tax in Russia as a promising tool for reduction of airborne emissions and development of road network. WIT Trans. Ecol. Environ. **198**, 391–402 (2015.) WIT Press, UK

Part IV
Energy Efficient Technologies

Improved Energy Efficiency and Environmental Safety of Transport Through the Application of Fuel Additives and Alternative Fuels

G. Genon, E. Magaril, R. Magaril, D. Panepinto, and F. Viggiano

1 Introduction

The global nature of anthropogenic challenges related to a rapid depletion of nonrenewable energy resources and related environmental pollution requires a new priority of economic development with the ecological imperative to prevent catastrophic consequences. The interdisciplinary nature of this problem requires the formation of a program of improving energy efficiency and environmental safety of transport, which is one of the main consumers of fossil fuels and emitters of toxic pollutants and greenhouse gases into the environment. Such a program should include a set of technological measures aimed at improving the quality of traditional motor fuels and the use of alternative fuels. The specific methods will differ depending on the specific conditions of the country considering the environmental-economic efficiency.

The Environment Action Programmes of the European Union [1] aim to achieve levels of air quality that do not result in unacceptable impacts on human health, and consequent initiatives have been adopted; but, despite these efforts, current air quality in Europe still leads to adverse impacts, such as exposure to particulate matter and ozone, damage to materials and cultural heritage, and impacts of persistent organic pollutants on human health. In Figs. 1 and 2 [2], an illustration is reported about the emission trends of selected pollutants and the contribution that different sectors make to emission of these pollutants. From these figures, it is clear

G. Genon · D. Panepinto · F. Viggiano
Politecnico di Torino, Turin, Italy

E. Magaril (✉)
Ural Federal University, Yekaterinburg, Russia

R. Magaril
Tyumen State Oil and Gas University, Tyumen, Russia

© Springer International Publishing AG, part of Springer Nature 2018
S. Syngellakis, C. Brebbia (eds.), *Challenges and Solutions in the Russian Energy Sector*, Innovation and Discovery in Russian Science and Engineering,
https://doi.org/10.1007/978-3-319-75702-5_18

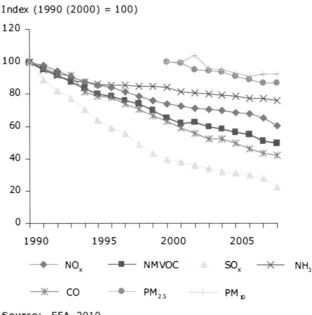

Index (1990 (2000) = 100)

Source: EEA, 2010.

Fig. 1 Emission trends for selected air pollutants, EU 27 [2]

that road transport, as concerns the air quality, is a significant contributor, besides other negative aspects, chiefly to tropospheric ozone precursor emissions.

As a consequence of these concerns, the European Union in the past 20 years has developed and implemented policies aimed at a cleaner vehicle fleet, by defining exhaust emission limits for new cars, developing abatement technologies, and implementing plans for modification of transport and movement paths in the cities.

In consideration of the fact that a large part of EU population [3] lives close to roads with a large year vehicle traffic and that standards for PM_{10}, $PM_{2.5}$, and NO_x are in many areas at the moment not attained (as it can be observed from Ref. [4]), it is clear that some other interventions are required, in order to obtain a higher sustainability in road transport.

On the other hand, transport is responsible for more than 20% of total greenhouse emissions in EU [5], on account of the fact that an increased transport volume leads to a higher effect in comparison with efforts directed to more efficient cars. Concerning this aspect, besides initiatives in direction of a higher car efficiency and a lower fuel consumption (in Fig. 3, from Ref. [6], an indication in this sense can be observed), it is clear that the general approach directed to nonfossil fuels and renewable forms of fuels and energy can have a positive and useful application, like in large combustion plant and in strategies for thermoelectric production, also in the scenarios of alternative, renewable, nonfossil automotive fuels.

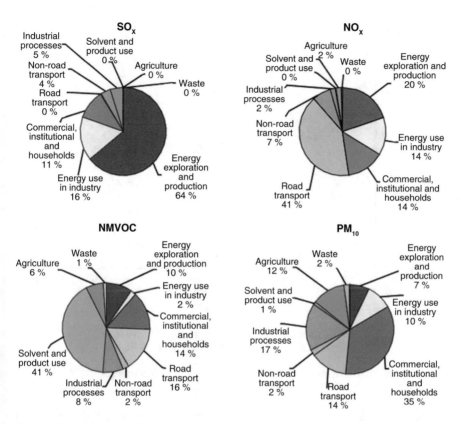

Fig. 2 The contribution made by different sectors to total emissions in 2008, EU 27 [2]

2 The Current State of Research on Improving the Transport Energy Efficiency and Sustainability

The basic areas of the current investigations in improving environmental safety and energy efficiency for the transportation facilities are an optimization of the structure and an improvement of the characteristics of the vehicle fleet (including an improvement in the design of engines and vehicles themselves and equipping cars with catalyst converters, etc.); an improvement of the maintenance system, road network, and traffic management, reducing the negative impact of climatic conditions; and an optimization of speed limits, improving the environmental characteristics of fuels, including an improvement of the quality of traditional fuels by refining methods, by the use of fuel additives and alternative fuels and energy sources.

The influence of various constructional and technological parameters on environmental and operational characteristics of transport was precisely investigated (including the works of Russian scientists Trofimenko [7], Erokhov [8]), and research in this area continues. Taking into account that any modernization in the design of engines and vehicles cannot be effective enough with low quality of fuels,

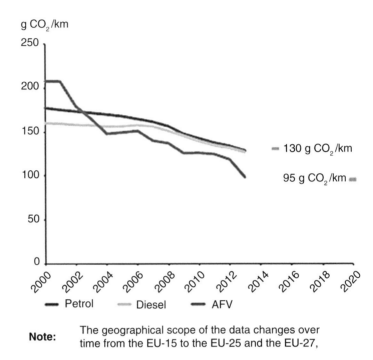

Fig. 3 Evolution of CO_2 emissions from new passenger cars by fuel type (EU 27) [6]

an improvement of the quality of conventional motor fuels and the search for possible alternatives are a priority. Many investigations are devoted to the fuel quality issue [9–14] and to the potential improvement in the quality of motor fuels by oil refining industry [15–18]. However, reformation of the refining industry requires huge investments and is acceptable only for rich countries.

One of the least expensive, fastest, and most efficient ways to influence the properties of gasoline and diesel fuels is the use of fuel additives, which provide the same or greater effect in comparison with the changes in the technology [12–14, 19–22]. All US refining companies, for example, introduce into the motor fuels produced additives, advertising the benefits of improved fuel quality,without revealing the additives' composition. The presence of hundreds of patents on additives for motor fuels and huge nomenclature of additives on the market demonstrates their great potential in correction of the motor fuel properties. Meanwhile, some of the proposed additives, improving one property of motor fuel while degrading others, become the reason of corrosion, gum formation, increased the cost of fuels, and so on and so forth. Apparently, the development of additives for motor fuels to improve their environmental properties does not acquire an adequate scale due to aggressive introduction of catalyst converters. Meanwhile, the use of catalyst converters increases fuel consumption; moreover, there are problems in their use at the low level of the fuel's quality and control.

It should be noted that research in the development of fuel additives are carried out mainly by large companies (such as ExxonMobil, Chevron, NewMarket Corp, Lubrizol Corp, Baker Hughes Inc., Shell, BASF AG, Honeywell Inc., The Dow Chemical Com, BP PLC); the authors of inventions are employees of companies and do not have rights to publish scientific papers, and the results of investigations are presented in the patents.

Accessing patent data indicates that different organizations have focused their research on various categories and areas of application of additives – detergent, antioxidant properties, and others. There is also some further research on catalysts. Chevron, General Electric Co, Massachusetts Inst., Shell, and other companies have progressed in the direction of the development of nano-additives to reduce engine wear.

In order to limit the exhaust emissions, many oxygenated additives have been proposed and a careful assessment of their efficiency has been made, but subsequently, in many cases, they cause secondary toxic emitted pollution [23]. Chiefly oxygenated compounds have been considered in this direction, in the first time with a large use of MTBE [24], now with higher interest in alcohols, chiefly ethanol and butanol.

The evaluation of effects of addition of these components on emissions, both regulated (particles, NO_x, CO) and unregulated substances (aromatics, aldehydes, unburned additives), has been studied, on different scale of application and with different ratios of introduction in fossil fuel of the improving agent.

For example, concerning the ethanol addition, a careful study has been conducted by Manzetti and Andersen [25], by individuating the effects of addition on over-indicated regulated and unregulated emissions and also the consequences from the catalytic efficiency of the post-combustion systems.

A decrease of exhaust emissions has been individuated by the addition of oxygenates, with emphasis on NO_x, CO, THC); for example, in [26] it has been determined that the addition of alcohols (methanol, ethanol, propanol from 5% to 20%) could decrease the NO_x until to 60%. On the contrary, the addition of oxygenated compounds affects the exhaust emission of some unregulated pollutants: this aspect can be observed for aldehydes from ethanol and organic acids formed from the additives precursors, while benzene emissions from engine can be considered as decreasing by introduction of oxygenates in the gasoline.

The potential effects arising from ethanol addition can be individuated by considering that blending gasoline and ethanol encompasses new components in the emissions [27], and in order to evaluate the potential toxicity of this qualitatively modified combustion, a complete degradation chart for oxygenated blends must be determined, starting from chemical conversion to atmospheric oxidation path. An example of this chart, from Ref. [28], is reported in Fig. 4.

As a conclusion at this point, the modification of gasoline composition very probably leads to advantages for conventional parameter emissions, to an extent that depends on many specific thermo-kinetic aspects, but careful attention must be devoted to secondary volatile pollutants arising from oxygenated precursors.

With the same approach, the objective of reduction of environmental impact from diesel fuel combustion led to many studies that consider addition of oxygenated

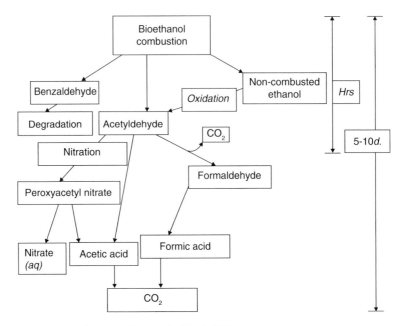

Fig. 4 Degradation charts for bioethanol – blends [28]

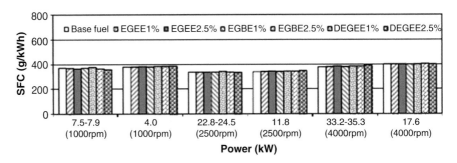

Fig. 5 Effect of EGEE, EGBE, and DEGEE addition on specific consumption [29]

compounds to diesel fuel. This addition could improve fuel combustion in engines, reducing pollutant emissions (CO, NO_x, unburned hydrocarbons, smoke) and at the same time increasing engine efficiency. Figure 5 (from Ref. [29]) is an example of the effect of the use of different glycol ethers on the principal pollutants.

3 Limitation of Climate Change

As substitution of conventional fossil fuels, biofuels are increasing their use on the market, on account of the possibility to exploit local low-cost resources but chiefly with the objective to reduce the GHG emissions. In fact, the CO_2 emission that is

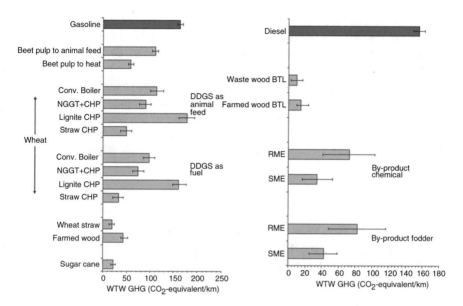

Fig. 6 Transport and environment 2007 [30]

created during the combustion is at least partially compensated from the CO_2 absorbed during the plant growth. The advantage in this sense, evaluated with a complete well-to-wheel analysis [30], by considering different biofuels, is reported in Fig. 6: the figure illustrates the total net emission of GHG needed to produce and consume enough fuel to move a vehicle for 1 km. In the figure there is also an indication of different raw materials for biofuels and different process options.

From the figure it is easy to observe a clear saving along most production pathways but also a large variation in the net saving depending on the chosen scenario.

In comparison with this clear positive effect, there are important concerns about the potential negative effects of biofuels during the growth of considered raw vegetal materials.

The rise of demand for biomass for energetic fuel production and use could put additional pressure on farmland and agricultural or forest biodiversity, and it could also affect the right use of natural resources as soil and natural water. It could very probably create negative aspects concerning waste minimization and environmentally oriented farming.

Another aspect that must be considered is the balance between the used farmland destined to biomass cultivation to support fuel production, and the use for food supply, and also the balance between biomass use for biofuels (that consumes an important quantity of energy) and biomass for bio-electricity or CHP that can be more energetically efficient.

Concerning this last point of global sustainability of a scenario of biofuels from natural materials, it is very important to consider the potential option of second-generation biofuels (obtained by using nonfood products and by using marginal territories); chiefly the exploitation of lignocellulosic substrates offers a very positive perspective, also if many problems of biological conversion paths and optimization of energy use must be solved.

Many studies, performed by using the LCA approach, evaluation of cost and benefits, definition of carbon footprint, and territorial best planning by using multi-criteria analysis, have been conducted concerning the possibility to extend the production of biofuels; also, obviously, the economic convenience must be considered, chiefly on account of the original territorial area where the original raw vegetal material is produced.

4 Improving the Vehicles' Energy Efficiency and Sustainability Through the Application of a Multifunctional Additive to Motor Fuels

By analyzing the data on physical and chemical processes in internal combustion engines, catalytic properties of substances, and tribological data, the formula of the multifunctional additive was theoretically proved, and the technology of its synthesis was developed [12–14, 21, 22].

An introduction of the multifunctional additive to traditional fuel in very small amounts (9.25 ppm for gasoline and 27.75 ppm for diesel fuel) provides a comprehensive positive impact on the properties of the fuel and also on the engine performance through a combination of a high surface activity of oil-soluble compound and the catalytic activity of the element in its composition.

Modifying of motor fuels being used by introduction of the developed fuel additive significantly improves energy efficiency and environmental performance of vehicles, and that is confirmed by the results of laboratory, bench, and road tests [21].

The additive is introduced in motor fuels in a controlled ultra-low concentration. Subsequent treatment of the engine by this author's technology, after decomposition of the additive in flame provides coating of the surfaces of internal combustion engine (ICE) by a catalytically active, in the gasification reactions, layer of metal in the nanocrystalline state, providing engine cleanliness. The formation of the nanolayer was confirmed by electron microscopic studies [31, 32].

Tests have shown that formation of the catalytic nanolayer in the internal combustion engines provides reduced emissions of soot precursors – polycyclic aromatic hydrocarbons, including benzo(a)pyrene by 95% (Fig. 7) – and eliminates carbon deposits itself, thus decreasing emissions of gaseous toxic substances (by 20–35%) and greenhouse gases.

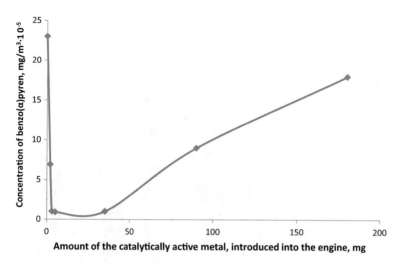

Fig. 7 The influence of modification of the gasoline by introducing the multifunctional additive on the presence of benzo(a)pyrene in the exhaust gases

Elimination of carbonization in the engine mitigates the temperature regime; this reduces the specific fuel consumption by 7–12% (Fig. 8) and the gasoline engines' requirements for the octane number of gasoline being used by up to 10 points [33].

Considering the annual motor fuel consumption, the obligatory application of fuels, modified by the additive introduction, should provide very significant economic and environmental effects and also reduces the requirements of the vehicle fleet for the high-octane gasolines.

A catalytic layer is cladding the engine surfaces, which reduces wear on the cylinder-piston group, and possesses a high corrosion resistance. Removal of soot from spark plugs of gasoline engines increases their lifetime.

It was established that even a single introduction of the additive within the fuel into the engine, followed by forming a catalytic layer on the working surfaces, leads to significant improvements in environmental and operational performance of gasoline engines. The catalytic layer retains its activity for a long enough time.

The constant use of the fuels, modified by the additive application, provides additional positive effects, improving detergent, lubricating properties, and reducing evaporation losses of gasoline [34]. It should be noted that atomic absorption analysis revealed that these modified fuels do not produce any additional toxic components in exhaust gases, in comparison with fuels before the additive application.

Considering the low cost of the proposed method to improve the energy efficiency and environmental performance of vehicles through using the motor fuels modified by the additive application, it is a prospect way both for developed and developing countries to solve the urgent task of rational use of fossil fuels and to reduce the transport negative impact on the environment.

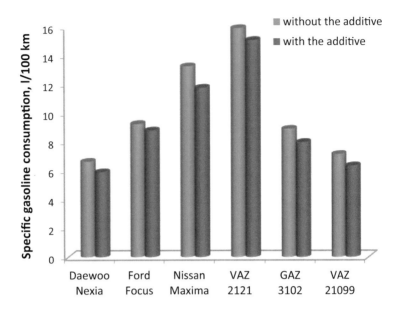

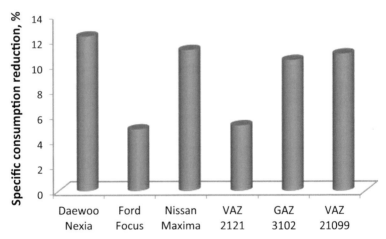

Fig. 8 The influence of gasoline modification by introducing the additive on its specific consumption

5   Conclusion

In order to improve the sustainability of road transport and automotive exploitation, there are some well-consolidated possibilities, chiefly corresponding to improvement in technological car devices, and some probably more performing but also more difficult to be adopted strategies, such as modification in mobility systems or territorial best planning. Besides these strategies, a careful consideration of the quality of fuels must be taken into account: some benefits from this could be produced concerning air quality arising from exhaust emissions, and chiefly important limitations in GHG emissions could be obtained. It is in any case very important to take into account all the aspects of fuel reformulation, with a global perspective, in order to balance the well identified advantages and to balance them in comparison with secondary pollution, territorial, and also social aspects of critical applicability.

An efficient method for improvement of energy efficiency and environmental safety of transport operation by application of the fuel additive is developed. Usage of the motor fuels, modified by introducing the additive, with minimal expenditure significantly improves the efficiency of vehicle operation, the engine state, and environmental characteristics.

Acknowledgment The work was supported by Act 211 Government of the Russian Federation, contract № 02.A03.21.0006.

References

1. EEA (European Environmental Agency) A closer look at urban transport TERM 2013: Transport indicators tracking progress towards environmental targets in Europe, EEA Report n. 11/2013
2. EEA (European Environmental Agency) Impact of selected policy measures on Europe's air quality, EEA Report n. 8/2010
3. ENTEC, Development of a methodology to assess population exposed to high levels of noise and air pollution close to major transport infrastructure, Final Report April 2006, Entec UK Limited
4. EEA (European Environmental Agency) Transport and environment: On the way to a new common transport policy TERM 2006: Indicators tracking transport and environment in the European Union, EEA Report n. 1/2007
5. EEA (European Environmental Agency) Greenhouse gas emission trends and projections in Europe 2006, EEA Report n. 9/2006
6. EEA (European Environmental Agency) Monitoring CO_2 emissions from passenger cars and vans in 2013, EEA Technical Report n. 19/2014
7. Trofimenko, Y.V.: Current problems of environmental engineering and supporting the Technosphere safety of motor complex [in Russian]. Safety Technos. **2**, 46–54 (2007)
8. Erokhov, V.I.: Petrol Injection: Design, Calculation, Diagnostics [in Russian], 552 p. Scientific and Technical Publishing House "Hot Line Telecom", Moscow (2011)
9. Azev, V.S., Emel'yanov, V.E., Turovskii, F.V.: Automotive gasolines. Long-term requirements for composition and properties. Khimiya i Tekhnologiya Topliv i Masel [in Russian]. **5**, 20–24 (2004)
10. Bauserman, J.W., Mushrush, G.W., Hardy, D.R.: Organic nitrogen compounds and fuel instability in middle distillate fuels. Ind. Eng. Chem. Res. **47**(9), 2867–2875 (2008)
11. Choi, B., Jiang, X., Kim, Y.K., Jung, G., Lee, C., Choi, I., Song, C.S.: Effect of diesel fuel blend with n-butanol on the emission of a turbocharged common rail direct injection diesel engine. Appl. Energy. **146**, 20–28 (2015)
12. Magaril, E.R.: Influence of the Quality of Engine Fuels on the Operation and Environmental Characteristics of Vehicles: Monograph [in Russian]. KDU, Moscow (2008)
13. Magaril, E.R., Magaril, R.Z.: Motor Fuels [in Russian], 2nd edn. KDU, Moscow (2010)

14. Magaril, E., Magaril, R.: Motor Fuels: The Problem of Energy Efficiency and Environmental Safety: Monograph [in Russian]. LAP LAMBERT Academic Publishing GmbH& Co, Saarbrücken (2012)
15. Shiriyazdanov, R.R., Rysaev, U.S., Nikolaev, E.A., Shiriyazdanov, R.M., Turanov, A.P., Akhmetov, S.A., Bulyukin, P.E.: Manufacturing of alkylate gasoline on the polycation-decationated form of nickel-and cobalt-promoted zeolite. Chem. Technol. Fuels Oils. **44**(5), 295–297 (2008)
16. Robinson, P.R.: Petroleum processing overview (Chapter 1). In: Hsu, C.S., Robinson, P.R. (eds.) Practical Advances in Petroleum Processing, vol. 1, pp. 1–78. Springer Science – Business Media, Inc., New York (2006)
17. Stanislaus, A., Marafi, A., Rana, M.S.: Recent advances in the science and technology of ultra low sulfur diesel (ULSD) production. Catal. Today. **153**(1–2), 1–68 (2010)
18. Magaril, E.: The solution to strategic problems in the oil refining industry as a factor for the sustainable development of automobile transport. WIT Trans. Ecol. Environ. **190**(2), 821–832 (2014.) WIT Press: UK
19. Danilov, A.M.: Fuel additives as a solution to chemmotological problems. Chem. Technol. Fuels Oils. **50**(5), 406–410 (2014)
20. Keskin, A., Gürü, M., Altiparmak, D.: Influence of metallic based fuel additives on performance and exhaust emissions of diesel engine. Energy Convers. Manag. **52**(1), 60–65 (2011)
21. Magaril, E.: Improving car environmental and operational characteristics using a multifunctional fuel additive. WIT Trans. Ecol. Environ. **147**, 373–384 (2011.) WIT Press: UK
22. Magaril, E.: The influence of carbonization elimination on the environmental safety and efficiency of vehicle operation. Int. J. Sustain. Dev. Plan. **8**(4), 1–15 (2013)
23. Song, C.L., Zhang, W.M., Pei, Y.Q., Fan, G.L., Xu, G.P.: Comparative effects of MTBE and ethanol additions into gasoline on exhaust emissions. Atmos. Environ. **40**, 1957–1970 (2006)
24. An, Y.-J., Kampbell, D.H., Sewell, G.W.: Water quality at five marinas in Lake Texoma as related to methyl tert-butyl ether (MTBE). Environ. Pollut. **118**, 331–336 (2008)
25. Manzetti, S., Andersen, O.: A review of emission products from bioethanol and its blends with gasoline. Background for new guidelines for emission control. Fuel. **140**, 293–301 (2015)
26. Zervas, E., Montagne, X., Lahaye, J.: Emissions of regulated pollutants from a spark ignition engine. Influence of fuel and air/fuel equivalence ratio. Environ. Sci. Technol. **37**, 3232–3238 (2003)
27. Lopez-Aparicio, S., Hak, C.: Evaluation of the use of bioethanol fuelled buses based on ambient air pollution screening and on-road measurements. Sci. Total Environ. **452**, 40–49 (2013)
28. Sauer, M.L., Ollis, D.F.: Photocatalyzed oxidation of ethanol and acetaldehyde in humidified air. J. Catal. **158**, 570–582 (1996)
29. Gómez-Cuenca, F., Gómez-Marín, M., Folgueras-Díaz, M.B.: Effects of ethylene glycol ethers on diesel fuel properties and emissions in a diesel engine. Energy Convers. Manag. **52**, 3027–3033 (2011)
30. JRC/Concawe/Eucar (2006). Available at: http://ies.jrc.ec.europa.eu/wtw.html. Last accessed on 27 Apr 2015
31. Magaril, E.: Improving the efficiency and environmental safety of gasoline engine operation. WIT Trans. Built Environ. **130**, 437–485 (2013.) WIT Press: UK
32. Magaril, E.: Carbon-free gasoline engine operation. Int. J. Sustain. Dev. Plan. **10**(1), 100–108 (2015)
33. Magaril, E.R., Magaril, R.Z., Bamburov, V.G.: Specific features of combustion in gasoline-driven internal combustion engines. Combust. Explos. Shock Waves. **50**(1), 75–79 (2014)
34. Magaril, E.R.: Reducing gasoline loss from evaporation by the introduction of a surface-active fuel additive. WIT Trans. Built Environ. **146**, 233–242 (2015.) WIT Press: UK

Structural and Electrical Properties of Composites Based on Ni and NiAl Alloys for SOFC Application

S. M. Pikalov, E. Yu. Pikalova, and V. G. Bamburov

1 Introduction

To date, in an effort to increase the efficiency of solid oxide fuel cells (SOFCs), thin film fabrication methods have been extensively used. With all conventional techniques such as tape casting, screen printing and deep coating, the necessary application of high temperatures leads to an appearance of various defects in the case of large area cells and remarkable interaction between the functional layers at the production stage. Along with high deposition rates of materials, the main advantages of the air plasma spraying (APS) method in applying for electrode-supported SOFC fabrication, excluding the firing processes, are ease of control of component composition and microstructure and the ability to produce cells in a variety of different shapes [1].

Ni-YSZ composite anodes possess advanced catalytic and electrical properties and are known to have good compatibility with solid-state electrolytes based on ZrO_2, CeO_2, $LaGaO_3$ and $BaCeO_3$ [2]. Regardless of their drawbacks and the numerous investigations searching for new SOFC anodes, Ni-YSZ still remains a benchmark material in this field due to a lack of alternative high-performance anodes. The matter of long-term stability of Ni-cermet is directly connected with

S. M. Pikalov
Institute of Metallurgy, UB RAS, Yekaterinburg, Russia

E. Y. Pikalova (✉)
Institute of High Temperature Electrochemistry, UB RAS, Yekaterinburg, Russia

Department of Environmental Economics, Ural Federal University, Yekaterinburg, Russia

V. G. Bamburov
Department of Environmental Economics, Ural Federal University, Yekaterinburg, Russia

Institute of Solid State Chemistry, UB RAS, Yekaterinburg, Russia

© Springer International Publishing AG, part of Springer Nature 2018
S. Syngellakis, C. Brebbia (eds.), *Challenges and Solutions in the Russian Energy Sector*, Innovation and Discovery in Russian Science and Engineering,
https://doi.org/10.1007/978-3-319-75702-5_19

the anode's structure, uniformity of the distribution of Ni particles (preferable nanosized) into the ceramic matrix and the prevention of their coarsening. In addition to particle size and the Ni/ceramic component ratio, another important influencing factor on the coarsening of Ni particles in cermet is wettability, and this is strongly affected by the composition of the ceramic component. In Ref. [3] the functional properties of NiO-ScSZ anodes were improved by using the additive oxides MgO and Al_2O_3. It was reported that small Ni particles formed during anode reduction were stable against coarsening. In Ref. [4] it was shown that small additives of Al_2O_3 considerably improve the long-term stability of Ni-scandia-stabilized zirconia (ScSZ) anodes. A small amount of $NiAl_2O_4$ was found to have formed at the interface of NiO and Al_2O_3 during the anode sintering, and, after reducing, Ni particles were fixed in the ScSZ ceramic matrix by the $NiAl_2O_4$ and the defect spinel phase of $NiAl_{10}O_{16}$. The composite configuration prevented aggregation of Ni particles under operating conditions and provided stability of the anode structure and excellent electrical properties. Interesting results were presented in Ref. [5], where a single SOFC with a thin film YSZ electrolyte formed on a supported anode on a base of NiAl strain-reinforced alloy demonstrated improved performance (500 mW/cm^2 at 700 °C) and stable anode resistance (6–10 × 10^{-3} Ω × cm at 600 °C in a gas mixture of Ar/2–3%H_2 for a minimum of 300 h).

Intermetallides of NiAl system are widely used in different high-temperature applications owing to their properties such as high melting temperature, high creep strength, high corrosion and oxidation resistance [6]. In the present work, Ni and NiAl alloys were used as the components of the composites on a base of YSZ or Al_2O_3 ceramic powders. The structural, electrical and thermal properties of composites sprayed using the APS method and thermally treated in different atmospheres were investigated in terms of their prospective usage as anodes in SOFC and other electrochemical devices.

2   Experimental Details

2.1   Feedstock Powders

As the feedstock powders for the production of the composite materials, flowable Ni powder of 20–80 μm fraction (Norilsk Nickel, Russia) or NiAl (Al-clad Ni with Al content of 5% ($Ni_{95}Al_5$) and 15% ($Ni_{85}Al_{15}$), Tulachermet, Russia) and α-Al_2O_3 (Al_2O_3 content of no less than 99%, RUSAL, Russia) or 9.5YSZ (ZrO_2 stabilized by 9.5 mol.% Y_2O_3, Titov's lab firm, Yekaterinburg, Russia) of regular spherical shape with a mean particle size of 50 μm and a flowing rate of 60 g/min were used. According to XRD (diffractometer DMAX-2500) 9.5YSZ has a cubic structure (Fm-3m space group, a = 5.1442 ± 0.0005 Å), and α-Al_2O_3 powder contains two phases with rhombohedral (R-3c space group, a = b = 4.7690 ± 0.0005 Å, c = 13.0220 ± 0.0033 Å) and monoclinic (P2 space group with lattice parameters a = 9.5558 ± 0.0078 Å, b = 5.1294 ± 0.0040 Å, c = 9.1620 ± 0.0103 Å) structures.

Al content of up to 5% in NiAl has no effect on the XRD pattern of Ni (only slightly shifting the peak to the left, a = 3.5249 ± 0.0003 Å). An increase in the Al content by up to 15% leads to the displacement of the main peak in the Ni diagram with the occurrence of reflexes of intermetallic compound AlNi$_3$ (Pm-3m, a = 3.5712 ± 0.0003 Å).

2.2 Preparation of Porous Samples of Tubular Shape by APS Method

Mechanical mixtures of the oxide powders Al$_2$O$_3$ or 9.5YSZ and Ni or nickel alloys NiAl in equal weight proportions were used for preparation of the sample of tubular shape by the air plasma spraying (APS) method [7]. The spraying was done by a semiautomatic APS industrial set using air plasma spray-type UMP-7 apparatus and plasma torches of our own design with an external rotating anode. The powder mixtures were deposited on a metal pin 9 mm in diameter with a precoated antiadhesive layer. After spraying, tubular-shaped samples (about 10 mm in diameter and 220 mm in length) were taken off the pin by soaking in water to remove the adhesive layer. The thickness of the tubes' wall was 400–600 μm (Fig. 1). After spraying the porosity of the samples was evaluated by the hydrostatic weighing method and was found to be equal to 20–22%.

The phase content of the samples was examined after deposition, after heat treatment for 2 h at 1350 °C in argon, and after the subsequent reducing in a hydrogen atmosphere for 2 h at 1350 °C. XRD study of the samples was performed using a DMAX-2500 with Ni-filtered CuK$_\alpha$ radiation in the range of 25° ≤ 2θ ≤ 85°. The surface morphology of the samples was studied by a scanning electron microscope Auriga Crossbeam Workstation (Carl Zeiss) combined with an INCA SEM MA System. The electrical conductivity of the samples was measured by the four-point

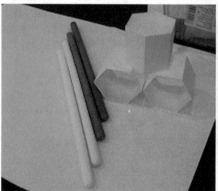

Fig. 1 Porous samples of tubular shape fabricated by APS method (white samples made from Al$_2$O$_3$, dark – from the cermets)

DC technique in hydrogen using a microprocessor system ZIRCONIA-318. Thermal expansion of samples in air and argon was studied using a dilatometer Tesatronic TT60.

3 Results and Discussion

3.1 Structure and Phase Changes After Heat Treatment

In as-sprayed Ni + 9.5YSZ samples, the oxide phase $Zr_{0.758}Y_{0.242}O_{1.879}$ with a cubic structure and a lattice parameter a = 5.1479 ± 0.0003 Å and metallic nickel (a = 3.5272 ± 0.0003 Å) were detected together with a small amount of 1.9 wt.% NiO (a = 4.1828 ± 0.0007 Å). The composition was stable during heat treatment in different atmospheres. After firing in argon with following reduction in hydrogen, NiO was completely reduced up to the metallic state (Fig. 2a).

Replacing nickel with $Ni_{95}Al_5$ is marked by a slight shift of the Ni and 9.5YSZ peaks to a smaller angle area and also by the appearance of a small amount of NiO and $NiAl_2O_4$. When treated in Ar, the nickel oxide was reduced to metallic nickel and the amount of spinel in the samples increased. The activation energy of the reduction reaction of NiO in the temperature 291–509 °C range is 18 kJ mol^{-1} and that of the reduction reaction of spinel in 1014–1264 °C range is 134 kJ mol^{-1} [8]. This means that NiO reduces more readily than $NiAl_2O_4$. After reduction in hydrogen, traces of Al_2O_3 were detected in the mixture of Ni and 9.5YSZ (Fig. 2b). In the $Ni_{85}Al_{15}$–9.5YSZ cermet samples, the formation of a remarkable amount of $NiAl_2O_4$ and $Al_{0.14}Ni_{0.86}$ (cubic structure Fm-3m, a = 3.5491 ± 0.0003 Å) was found after spraying. Treatment in argon led to the appearance of non-stoichiometric spinel which, after reduction in hydrogen, transformed into metal nickel and alumina (Fig. 2c).

After being held in argon, the samples changed in colour from grey to green and also significantly increased in size (by 7.4%) which was not observed in the other compositions based on 9.5YSZ. After reducing in hydrogen, the samples' sizes differed from their original state by only 0.3%. During heating in hydrogen, the changes in sample sizes were less remarkable.

In the case of Ni-Al_2O_3, after spraying the main phases were metallic Ni, γ-Al_2O_3 phase with a cubic structure (space group Fd-3 m, lattice parameter a = 7.9105 ± 0.0006 Å) and the nuclei of α-phase crystallization (Fig. 3a). After annealing in argon, the γ-Al_2O_3 phase transformed into α-Al_2O_3. It should be noted that after spraying in the Al_2O_3-cermet Ni was found mainly in a metallic state, only in the sample Ni-Al_2O_3 was the NiO content found to be about 0.5%.

With the substitution of Ni to NiAl alloys, non-stoichiometric spinel was detected (Fig. 3b, c) in addition to main phases. After Ar treatment it transformed into spinel $NiAl_2O_4$ and after annealing in hydrogen decomposed to Ni and α-Al_2O_3 according to the scheme:

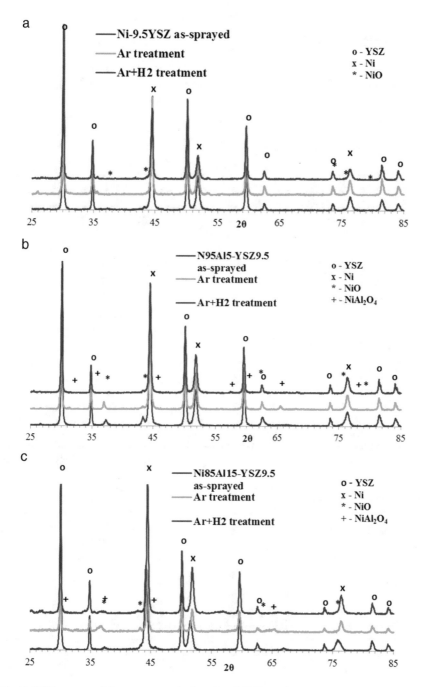

Fig. 2 XRD patterns of the cermet samples based on 9.5YSZ with Ni (**a**) and NiAl alloys $Ni_{95}Al_5$ (**b**) and $Ni_{85}Al_{15}$ (**c**)

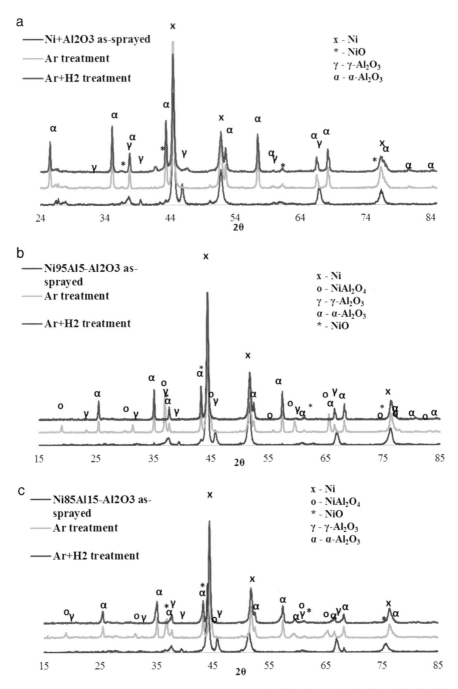

Fig. 3 XRD patterns of the cermet samples based on Al_2O_3 with Ni (**a**) and NiAl alloys $Ni_{95}Al_5$ (**b**) and $Ni_{85}Al_{15}$ (**c**)

$$\text{Ni}_2\text{Al}_{18}\text{O}_{29} \rightarrow\ ^{\text{T,Ar}} \rightarrow \text{NiAl}_2\text{O}_4 \rightarrow\ ^{\text{T,H}_2} \rightarrow \text{Ni} + \text{Al}_2\text{O}_3 + \text{H}_2\text{O} \uparrow .$$

The colour of all Al_2O_3-based samples changed from grey to green under an Ar atmosphere and then again became grey after reducing in H_2. The most significant change in size among the composites based on alumina was found for $\text{Ni}_{85}\text{Al}_{15}$-$\text{Al}_2\text{O}_3$ after Ar treatment, but it was less than that in the case of $\text{Ni}_{85}\text{Al}_{15}$–9.5YSZ.

3.2 Electrical Properties of the Cermets

Electrical properties of the cermets were investigated in wet hydrogen on the samples after treatment in Ar with following reducing in H_2. The conductivity of anode cermets with Ni and $\text{Ni}_{85}\text{Al}_{15}$ measured in wet hydrogen (3% H_2O) ranges from 100 to 150 S/cm at the temperatures of 600–900 °C.

The doping of Ni with Al in the range of solubility (5 mol%) leads to a significant increase in conductivity. It shows good electrocatalytic activity for hydrogen oxidation reactions. The maximal value was found to be 1364 S/cm for $\text{Ni}_{95}\text{Al}_5$-$\text{Al}_2\text{O}_3$ cermet at 600 °C, which is in the top range of the values of conductivity of the usual Ni-cermet anodes presented in the literature [9].

After reducing Ni-9.5YSZ in a hydrogen atmosphere, it was found that the Ni in the ceramic matrix formed into relatively large isolated particles, and this could be a possible reason for their reduced contact and conductivity (Fig. 4a, b). The substitution of Ni by nickel alloy in YSZ-based cermet and the use of Al_2O_3 as a ceramic base prior to heating in an oxidizing atmosphere led to the formation of thin barriers of Al_2O_3 and/or NiAl_2O_4 on the surface of the deposited layers. In a reducing atmosphere, this films break up into fine Ni and Al_2O_3 preventing the consolidation of nickel particles (Fig. 4c, d). An increase in Al content of up to 15% in the NiAl alloy above the solubility limit leads to the appearance of nonreversible compounds such as AlNi_3 which may be the reason for a further reduction in conductivity.

3.3 Thermal Expansion

The dependence of the thermal expansion of Ni-9.5YSZ during heating in air and in Ar was non-linear with a sharp increase in the high-temperature interval. The calculated CTE values were 8.4×10^{-6} K^{-1} (25–630°C), 31.3×10^{-6} K^{-1} (630–730°C) and 58.6×10^{-6} K^{-1} (730–900°C), respectively. The same extreme expansion was observed for all samples with YSZ ceramic matrix regardless of the metallic component. It is probably connected with Ni oxidation by residual oxygen in commercial argon. However, in our investigations, Al_2O_3-based cermets demonstrated even expansion under the same conditions across all temperature ranges. Calculated CTE values of $\text{Ni}_{95}\text{Al}_5$-$\text{Al}_2\text{O}_3$ change smoothly from 9.1 to 11.1×10^{-6} K^{-1} in the range of 25–900°C.

Fig. 4 Microstructure of cermets after spraying and after reducing in H_2: (**a**, **b**) Ni-9.5YSZ; (**c**, **d**) $Ni_{95}Al_5$-Al_2O_3

4 Conclusions

In order to form porous composite anodes of tubular shape by the air plasma spraying method, mechanical mixtures of the oxide powders Al_2O_3 or 9.5 YSZ and metal alloys $Ni_{100-x}Al_x$ (x = 0; 5; 15) were utilized in equal weight proportions. It was found that, after spraying, Ni and its alloys remain mainly in a metallic state. When substituting Ni with nickel alloys and using Al_2O_3 as a ceramic base under heating in an oxidizing atmosphere, a thin protective layer of Al_2O_3 and/or $NiAl_2O_4$ formed on the surface of deposited particles. In a reducing atmosphere, this film breaks up into fine Ni and alumina preventing the consolidation of nickel particles. The positive effect of such substitution is an increase in conductivity (1364 S/cm and 644 S/cm at 600 °C in hydrogen for Al_2O_3 + $Ni_{95}Al_5$ and 9.5YSZ + $Ni_{95}Al_5$, respectively) and reduced expansion of the samples. During heating in argon, the CTE of Al_2O_3 + $Ni_{95}Al_5$ changed smoothly from 9.1 to 11.1 × 10^{-6} K^{-1} in the range 25–900°C. The cermet having Al_2O_3 + $Ni_{95}Al_5$ content is a preferred candidate among the investigated cermets for further usage as a supporting anode in SOFC production.

Acknowledgements The present work was supported by Act 211 Government of the Russian Federation, contract № 02.A03.21.0006.

References

1. Hui, R., Wang, Z., Kesler, O., Rose, L., Jankovic, J., Yick, S., Maric, R., Ghosh, D.: Thermal plasma spraying for SOFCs: Applications, potential advantages, and challenges, review. J. Power Sources. **170**, 308–323 (2007)
2. Cowin, P.I., Petit, C.T.G., Lan, R., Irvine, J.T.S., Tao, S.: Recent progress in the development of anode materials for solid oxide fuel cells. Adv. Energy Mater. **1**(3), 314–332 (2011)
3. Shiratori, Y., Teraoka, Y., Sasaki, K.: $Ni_{1-x-y}Mg_xAl_yO$–ScSZ anodes for solid oxide fuel cells. Solid State Ionics. **177**(15–16), 1371–1380 (2006)
4. Orui, H., Chiba, R., Nozawa, K., Arai, H., Kanno, R.: High temperature stability of alumina containing nickel-zirconia cermets for solid oxide fuel cell anodes. J. Power Sources. **238**, 74–80 (2013)
5. Sadykov, V.A., Usoltsev, V.V., Fedorova, Y.E., et al.: Design of medium–temperature solid oxide fuel cells on porous supports of deformation strengthened Ni–Al alloy. Russian J. Electrochem. **47**(4), 488–493 (2011)
6. Pikalov, S.M., Polukhin, V.A., Kuznetsov, I.A.: Correlation of electromagnetic and mechanical properties of functional plasma sprayed coatings and criterion of non-destructive control quality. Metally. **6**, 146–152 (1995)
7. Pikalov, S.M., Selivaniv, E.N., Chumarev, V.M., Pikalova, E.Yu., Zaikov, Yu.P., Ermakov, A. V.: Composite electrode material for electrochemical devices. Patent Ru № 2523550, 20.07.2014
8. Sridhar, S., Sichen, D.U., Seetharaman, S.: Investigation of the kinetics of reduction of nickel oxide and nickel aluminate by hydrogen. Z. Met. **85**, 616–620 (1994)
9. Tsipis, E.V., Kharton, V.V.: Electrode materials and reaction mechanisms in solid oxide fuel cells: A brief review. J. Solid State Electrochem. **12**(11), 1367–1391 (2008)

Development of the Cathode Materials for Intermediate-Temperature SOFCs Based on Proton-Conducting Electrolytes

D. A. Medvedev and E. Yu. Pikalova

1 Introduction

The world is facing global climate change, a situation which calls for an effective low-carbon policy and efficient energy technologies. Energy technologies will also be crucial to achieving and maintaining a secure world energy supply. It is already evident that the use of standard technologies to produce electricity based on fossil fuels cannot satisfy the ever-growing demand for energy. The future for world energetic systems lies in the implementation of efficient and environmentally friendly technologies to produce electricity. From this perspective, hydrogen energy and fuel cells represent key technologies in attaining the renewable energy and emission reduction goals set worldwide [1].

Among the various kinds of fuel cells, solid oxide fuel cells (SOFCs) are advantageous because they are highly efficient at energy conversion and possess excellent fuel flexibility [2]. Principally, SOFCs can be divided into two main groups: the $SOFC(O^{2-})$s based on oxygen-conducting electrolytes (doped ZrO_2, CeO_2, Bi_2O_3, and $LaGaO_3$) and the $SOFC(H^+)$s based on proton-conducting electrolytes (doped $BaCe(Zr)O_3$, $LaNbO_4$, $Ba_2In_2O_5$). High-temperature proton-conducting oxide materials laying in a base of $SOFC(H^+)$s are of great fundamental interest because of the phenomenon of proton conductivity which appears along with oxygen-ionic conductivity in a humidified atmosphere. The practical interest connected with the use of such Co-ionic electrolyte materials in intermediate-temperature solid oxide fuel cells (IT-SOFCs) stems from the increased efficiency due to higher open circuit voltage and, correspondingly, the power output characteristics in comparison with SOFCs based on unipolar oxygen-ion conducting

D. A. Medvedev (✉) · E. Y. Pikalova
Institute of High Temperature Electrochemistry, UB RAS, Yekaterinburg, Russia

Ural Federal University, Yekaterinburg, Russia

© Springer International Publishing AG, part of Springer Nature 2018
S. Syngellakis, C. Brebbia (eds.), *Challenges and Solutions in the Russian Energy Sector*, Innovation and Discovery in Russian Science and Engineering,
https://doi.org/10.1007/978-3-319-75702-5_20

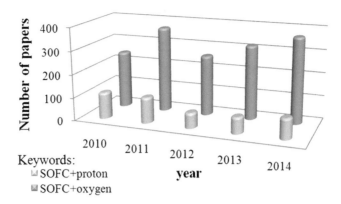

Fig. 1 Publication dynamics on the different search keywords in the SOFC field during the last 5 years (presented according to Scopus database)

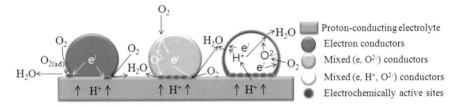

Fig. 2 Possible paths for the electrochemical reactions in cathode systems possessing different transport properties in contact with a proton-conducting electrolyte. The examples of electron conductors are Pt, and $LaNi_{0.6}Fe_{0.4}O_{3-\delta}$ [3], mixed (e',O^{2-}) conductor is $GdBaCo_2O_{5+\delta}$ [3], and mixed (e',O^{2-},H^+) conductor is $NdBa_{0.5}Sr_{0.5}Co_{1.5}Fe_{0.5}O_{5+\delta}$ [7]

electrolytes. Although SOFCs with protonic conductors are characterized by having the highest total efficiency (can exceed 80%) on account of thermodynamic and kinetic factors [3], they have been only moderately investigated compared to the $SOFC(O^{2-})$s (Fig. 1). Insofar as we know, there exists no commercial production of $SOFC(H^+)$s, one of the main reasons being the ongoing search for cathode materials that have all the necessary qualities for successful applications in contact with the optimal proton-conducting materials. To date there is much activity to enhance the SOFC's electrochemical characteristics by the development of new cathode materials which possess excellent electrocatalytic activity.

The first studies concerning $SOFC(H^+)$ based on $BaCe(Zr)O_3$ reported the use of a platinum cathode [3]. However, Pt is not preferred for practical applications due to its high cost. Moreover, the Pt electrodes were found to show significant polarization losses caused by the limited number of reaction sites at the cathode/electrolyte interface (Fig. 2). From this viewpoint, cathode materials with mixed oxygen-ionic and electronic conductivity widely used in $SOFC(O^{2-})$s so far as their application permits to significantly broaden the zone of electrochemical reaction (Fig. 2) [4] can be also applied in $SOFC(H^+)$s with Co-ionic conductors. However the development of an electrode material with proton conductivity would still be preferable since such cathodes allow the simultaneous transport of ionic (proton, oxygen-ion) and electronic defects under typical fuel cell operating conditions, thus offering the potential to extend the active sites for oxygen and proton reactions and, correspondingly, decrease the polarization losses. Additionally, the electrodes must be thermodynamically stable under working conditions: $400–900\ ^\circ C$, $10^{-5} \leq pO_2/atm \leq 0.21$, in the presence of H_2O and CO_2. Thermal affinity between electrolyte and cathode materials should be considered in order to attain both long-term stability and cycling.

Currently, there are many complex oxides that have been reported as having appropriate air electrode materials for $SOFC(H^+)$. Regardless of the fact that very high short-term power densities ($\sim 0.4\ W\ cm^{-2}$ at $600\ ^\circ C$, $\sim 1.0\ W\ cm^{-2}$ at $700\ ^\circ C$) have been obtained recently for such an $SOFC(H^+)$ [5–7], the issues associated with thermal affinity and chemical compatibility between electrolyte and cathode materials have not been precisely investigated. In this work the structural, electrical, and thermal properties of simple and layered cobaltites ($GdBaCo_2O_{5+\delta}$, $NdBaCo_2O_{5+\delta}$, $Ba_{0.5}Sr_{0.5}CoO_{3-\delta}$, $Y_{0.8}Ca_{0.2}BaCo_4O_{7+\delta}$), cobaltite-ferrites ($NdBa_{0.5}Sr_{0.5}Co_{1.5}$Fe_{0.5}O_{5+\delta}$, $GdBaCoFeO_{5+\delta}$, $Ba_{0.5}Sr_{0.5}Co_{0.8}Fe_{0.2}O_{3-\delta}$, $Ba_{0.5}Sr_{0.5}Co_{0.2}Fe_{0.8}O_{3-\delta}$, $La_{0.6}Sr_{0.4}Co_{0.2}Fe_{0.8}O_3$), nickelite ($La_2NiO_{4+\delta}$), and nickelate ($LaNi_{0.6}Fe_{0.4}O_{3-\delta}$) were investigated in terms of their perspective applications in intermediate-temperature SOFCs in contact with the $BaCe_{0.8}Y_{0.2}O_{3-\delta}$ and $BaZr_{0.8}Y_{0.2}O_{3-\delta}$ proton-conducting electrolytes with special attention given to their chemical compatibility.

2 Experimental Details

In the present study the cathode materials, except for $LaNi_{0.6}Fe_{0.4}O_{3-\delta}$ (LNF), were successfully synthesized according to the solid-state reaction method using oxides and carbonates of corresponding metals with a purity of not less than 99% as precursors. The citrate-nitrate combustion method was used to produce a single-phase LNF product. The designation of these materials and temperatures of the final synthesis T_{synt} and sintering T_{sint} are presented in Table 1. After calcination and ball milling, the powders were dry-pressed into disks or bar-shaped samples at a pressure of about 150 MPa and sintered at individually adjusted temperatures T_{sint}, which exceeded those of the final synthesis by $50–150\ ^\circ C$. The densities of the sintered specimens, determined from their geometrical size and weight, varied in the range of 85–95% of the theoretical value, calculated from the XRD data.

The electrolyte powders $BaCe_{0.8}Y_{0.2}O_{3-\delta}$ and $BaZr_{0.8}Y_{0.2}O_{3-\delta}$ were obtained by the citrate-nitrate combustion method. The precursors obtained were calcined at $1150\ ^\circ C$ (5 h) and finally sintered at $1450\ ^\circ C$ (5 h).

The synthesized powders' characteristics were determined by X-ray diffraction analysis (XRD, D/MAX-2200 Rigaku) using CuKα1 radiation. The identification of the materials' phase composition and crystal structure was performed by employing MDI Jade 6 software. The results of XRD analysis revealed the formation of single-phase structures for all the samples investigated, indicating the formation

Table 1 Chemical composition of the cathode materials and their temperatures of synthesis (T_{synth}) and sintering (T_{sint}) and soaking time (τ)

Composition	Designation	T_{synt} (°C)/ τ (h)	T_{sint} (°C)/ τ (h)	Space group
$Ba_{0.5}Sr_{0.5}CoO_{3-\delta}$	BSC	1050/5	1100/5	P63/mmc
$GdBaCo_2O_{5+\delta}$	GBC	1050/5	1150/5	Pmmm
$NdBaCo_2O_{5+\delta}$	NBC	1050/5	1150/5	P4/mmm
$Y_{0.8}Ca_{0.2}BaCo_4O_{7+\delta}$	YCBC	1050/5	1150/5	Cmc21
$Ba_{0.5}Sr_{0.5}Co_{0.2}Fe_{0.8}O_{3-\delta}$	BSCF28	1150/5	1200/5	Pm3m
$Ba_{0.5}Sr_{0.5}Co_{0.8}Fe_{0.2}O_{3-\delta}$	BSCF82	1100/5	1150/5	Pm3m
$La_{0.6}Sr_{0.4}Co_{0.2}Fe_{0.8}O_{3-\delta}$	LSCF	1150/5	1200/5	R3C
$GdBaCoFeO_{5+\delta}$	GBCF	1100/5	1150/5	P4/mmm
$NdBa_{0.5}Sr_{0.5}Co_{1.5}Fe_{0.5}O_{5+\delta}$	NBSCF	1100/5	1200/5	Pm3m
$Ba_{0.5}Sr_{0.5}FeO_{3-\delta}$	BSF	1200/10	1350/5	Pm3m
$La_{0.75}Sr_{0.2}MnO_{3-\delta}$	LSM	1200/10	1350/5	R3C
$La_2NiO_{4+\delta}$	LN	1300/10	1450/5	Fmmm
$LaNi_{0.6}Fe_{0.4}O_{3-\delta}$	LNF	1300/10	1450/5	R3C

of solid-state solutions. Corresponding data on crystal symmetry are listed in Table 1.

In order to estimate the chemical compatibility, the powdered cathode materials were thoroughly mixed with electrolyte powders in the weight ratio 1:1, calcined at 1100 °C for 10 h, and then analyzed by XRD. The calcination temperature was selected based on the literature data which show that the cathode functional layers usually form on the electrolyte surface at 1100 °C.

The thermal expansion of the materials was carried out using a Tesatronic TT-80 dilatometer between room temperature and 900 °C with a heating/cooling rate of $3\,°C\,min^{-1}$ in an air atmosphere. The average thermal expansion coefficients (TECs) were found from the fitting of the linear region of the dilatometric curves.

The conductivity of the samples was investigated at 500–900 °C in air by a standard four-probe dc method utilizing the microprocessor system ZIRCONIA-318.

3   Results and Discussion

3.1   Chemical Compatibility

After calcination at 1100 °C for 10 h, the XRD analysis was used to determine the characteristics of the powders' mixtures of the electrode and electrolyte materials and to estimate the degree of chemical interaction between them. It was found that the BCY electrolyte possesses a low chemical stability in contact with most of the cathode compositions. Although the structures of the main phases in

Table 2 Summary results of chemical and thermal compatibilities and electrical properties of the investigated cathode materials. The gray color of cells indicates the appropriate properties. The chemical expansion is marked by * symbol

Designation	Impurity phase(s)		$\alpha \cdot 10^6$ (K^{-1})	σ (S cm^{-1})	
	BCY	BZY		600 (°C)	700 (°C)
BSC	CeO$_2$, SrO	No interaction	15.6	420	385
GBC	CeO$_2$, BaCoO$_x$, BaO	No interaction	21.3	430	330
NBC	BaCoO$_x$, BaCoNd$_2$O$_7$	YBa$_2$Fe$_3$O$_8$	23.1*	925	795
YCBC	CeO$_2$, YBaCo$_2$O$_5$, YBa$_2$Co$_3$O$_9$	No interaction	9.6	90	105
BSCF28	No interaction	No interaction	26.0*	15	11
BSCF82	CeO$_2$, SrO, BaO	YBa$_2$Fe$_3$O$_8$, BaO, Fe$_2$O$_3$	16.6*	580	530
LSCF	La$_2$CoO$_4$	YBa$_2$Fe$_3$O$_8$	19.9*	205	190
GBCF	CeO$_2$, BaCoO$_x$, Fe$_3$O$_4$, BaGd$_2$FeO$_7$	No interaction	17.2	40	35
NBSCF	BaCoO$_x$, Sr(Fe,Co)O$_3$	No interaction	26.9*	360	305
BSF	No interaction	No interaction	34.1*	11	7
LSM	CeO$_2$, YMn$_2$O$_5$	Mn$_3$O$_4$	10.7	140	115
LN	No interaction	No interaction	13.1	70	60
LNF	No interaction	No interaction	14.5	515	500

electrolyte/cathode mixtures remained the same as before treatment, different impurities and phases of interaction appeared (Table 2):

- CeO$_2$ for BSC, GBC, YCBC, BSCF82, GBCF, and LSM.
- SrO for BSC and BSCF82.
- Barium cobaltites for GBC, NBC, GBCF, and NBSCF and strontium cobaltite for LSCF.
- BaO and SrO for GBC and BSCF82, respectively.
- BaCoNd$_2$O$_7$ for NBC, La$_2$CoO$_4$ for LSCF, Fe$_3$O$_4$ for GBCF, Sr(Fe,Co)O$_3$ for NBSCF, YMn$_2$O$_5$ for LSM, and BaGd$_2$FeO$_7$ for GBCF.

The Y$_{0.8}$Ca$_{0.2}$BaCo$_4$O$_{7 + \delta}$ layered perovskite phase completely decomposes in the mixture with cerate after a treatment for 10 h at 1100 °C: YBaCo$_2$O$_5$, YBa$_2$Co$_3$O$_9$, and CeO$_2$ phases are fixed along with the main BaCeO$_3$ structure.

No significant interaction was observed in the calcined mixtures of BCY with BSCF28, BSF, LN, and LNF.

The obtained data confirm previously presented results for Co-containing electrode materials [8], showing an active Co-diffusion from cobaltites into BaCeO$_3$-based electrolytes and, as a consequence, the formation of BaO, BaCoO$_2$, and BaCoO$_3$

impurities. The decrease in Co-ions concentration results in some suppression of the impurity phase formation in the sequence $Ba_{0.5}Sr_{0.5}CoO_{3-\delta} - Ba_{0.5}Sr_{0.5}Co_{0.8}Fe_{0.2}O_{3-\delta} - Ba_{0.5}Sr_{0.5}Co_{0.2}Fe_{0.8}O_{3-\delta} - Ba_{0.5}Sr_{0.5}FeO_{3-\delta}$. For example, no impurity phases were detected in the mixture of BSF and BCY. The chemical activity of the cerate electrolyte with double cobaltites seems to be higher than that with simple cobaltite since formation of secondary phases was detected for all the calcined mixtures, containing $LnBaCo_2O_{5+\delta}$-based oxides.

$BaZrO_3$-based electrolyte being the material with higher tolerance factor (t) shows better chemical stability than the barium cerate in agreement with the concept of Goldschmidt [9]. As in the case of barium cerate, an interaction of the barium zirconate with cobaltites (NBC and LSCF) was detected and also with manganite (LSM) but to a lesser degree (trace amount of $YBa_2Fe_3O_8$ and Mn_3O_4 were detected, respectively). A high degree of chemical interaction was observed for the BSCF82 in contact with BZY material with the appearance of some secondary phases.

3.2   Thermal Compatibility

Investigation of the thermal properties of the cathode materials is critical in terms of evaluating their adhesion and the quality of the interface contact and, correspondingly, maintaining the mechanical stability of the electrolyte/electrode assembly during the formation and operation of the SOFCs.

However, studies have revealed that materials which possess excellent catalytic activity and transport properties such as simple and layered cobaltites show the TEC values to be more than $20 \cdot 10^{-6}$ K^{-1} which is 1.5–3 times higher than those measured for BCY and BZY which are $8–12 \cdot 10^{-6}$ K^{-1} (Table 2). It makes their application in contact with the BCY and BZY electrolytes questionable.

For the cobaltites, the strategy of partial substitution of Co-ions by other transition metals (M = Fe, Ni, Cu) or of the cation on A-site of Co-based perovskite by a cation with a lower ionic radius is usually adopted in order to decrease TEC values [3, 4, 11]. The comparison of thermal data for $GdBaCo_2O_{5+\delta}$ ($21.3 \cdot 10^{-6}$ K^{-1}) and $GdBaCoFeO_{5+\delta}$ ($17.2 \cdot 10^{-6}$ K^{-1}) confirms the proposed strategy. However, for many cobaltite-ferrites (LSCF, BSF, BSCF28, BSCF82, NBSCF), the thermal expansion behavior deviates from linearity resulting in the intense expansion of ceramics in the high-temperature range in comparison with that in the low-temperature range. This can probably be attributed to the unwanted chemical expansion caused by the presence of elements in different oxidation and spin states [10].

3.3 Electrical Compatibility

Conductivity of the cathode materials is one of the most important properties necessary for optimum performance with properties such as low polarization, serial, and in-plane resistances [4, 11]. As long as the porosity of the electrode layers after formation and co-sintering with the electrolyte surface averages 25–40%, the conductivity of the compact electrode samples should be no less than 100 S/cm. The conductivity of the cathode materials decreases when increasing the temperature in the interval of 500–900 °C which indicates its metallic type. Most of the developed materials possess an acceptable conductivity at 600 and 700 °C, except BSCF28, BSF, and GBCF (Table 2). The conductivity for LN and YCBC samples registered intermediate values. Therefore their application is preferable in combination with conductors that have high electronic conductivity.

4 Conclusion

In the present work, different cathode materials were successfully prepared and their structural, thermal, and electrical properties and chemical compatibility with the electrolytes were thoroughly investigated to find the appropriate compositions for their eventual application in $SOFC(H^+)$ with $BaCeO_3$ and $BaZrO_3$ proton-conducting electrolytes.

Analysis of the experimental data has shown that, if taking into account all the abovementioned properties, $LaNi_{0.6}Fe_{0.4}O_{3-\delta}$ and $La_2NiO_{4+\delta}$ electrode materials are the most suitable as the cathodes in contact with $BaCe_{0.8}Y_{0.2}O_{3-\delta}$ and $BaZr_{0.8}Y_{0.2}O_{3-\delta}$ proton-conducting electrolytes for intermediate-temperature SOFCs.

The layered $Y_{0.8}Ca_{0.2}BaCo_4O_{7+\delta}$ cobaltite has the closest TEC value with those for cerate and zirconate ceramics; however, it completely decomposes after a treatment of $Y_{0.8}Ca_{0.2}BaCo_4O_{7+\delta}/BaCe_{0.8}Y_{0.2}O_{3-\delta}$ mixture and can be recommended only for usage in contact with $BaZr_{0.8}Y_{0.2}O_{3-\delta}$ electrolyte.

Acknowledgments This work is supported by the Russian Foundation for Basic Research (grant № 13-03-00065); Program of UD RAS (project № 15-20-3-15), by Act 211 Government of the Russian Federation (contract № 02.A03.21.0006); and the Council of the President of the Russian Federation (grant № СП-1885.2015.1).

References

1. Fuel Cell and Hydrogen technologies in Europe-Financial and technology outlook on the European sector ambition 2014–2020, New Energy World. http://www.new-ig.eu/hydrogen-fuel-cells

2. Mahato, N., Banerjee, A., Gupta, A., Omar, S., Balan, K.: Progress in material selection for solid oxide fuel cell technology: A review. Prog. Mater. Sci. **72**, 141–337 (2015)
3. Medvedev, D., Murashkina, A., Pikalova, E., Demin, A., Podias, A., Tsiakaras, P.: $BaCeO_3$: Materials development, properties and application. Prog. Mater. Sci. **60**, 72–129 (2014)
4. Tsipis, E.V., Kharton, V.V.: Electrode materials and reaction mechanisms in solid oxide fuel cells: A brief review. I. Performance-determining factors. J. Solid State Electrochem. **12**, 1039–1060 (2008)
5. Min, S.H., Song, R.-H., Lee, J.G., Park, M.-G., Ryu, K.H., Jeon, Y.-K., Shul, Y.-G.: Fabrication of anode-supported tubular $Ba(Zr_{0.1}Ce_{0.7}Y_{0.2})O_{3-\delta}$ cell for intermediate temperature solid oxide fuel cells. Ceram. Int. **40**, 1513–1518 (2014)
6. Nien, S.H., Hsu, C.S., Chang, C.L., Hwang, B.H.: Preparation of $BaZr_{0.1}Ce_{0.7}Y_{0.2}O_{3-\delta}$ based solid fuel cells with anode functional layers by tape casting. Fuel Cells. **11**, 178–183 (2011)
7. Kim, J., Sengodan, S., Kwon, G., Ding, D., Shin, J., Liu, M., Kim, G.: Triple-conducting layered perovskites as cathode materials for proton-conducting solid oxide fuel cells. ChemSusChem. **7**, 2811–2815 (2014)
8. Lin, Y., Ran, R., Zhang, C., Cai, R., Shao, Z.: Performance of $PrBaCo_2O_{5+\delta}$ as a proton-conducting solid-oxide fuel cell cathode. J. Phys. Chem. A. **114**, 3764–3772 (2010)
9. Sammells, A.F., Cook, R.L., White, J.H., Osborne, J.J., MacDuff, R.C.: Rational selection of advanced solid electrolytes for intermediate temperature fuel cells. Solid State Ionics. **52**, 111–123 (1992)
10. Yaremchenko, A.A., Mikhalev, S.M., Kravchenko, E.S., Frade, J.R.: Thermochemical expansion of mixed-conducting $(Ba,Sr)Co_{0.8}Fe_{0.2}O_{3-\delta}$ ceramics. J. Eur. Ceram. Soc. **34**, 703–715 (2014)
11. Tucker, M.C., Cheng, L., DeJonghe, L.C.: Selection of cathode contact materials for solid oxide fuel cells. J. Power Sources. **196**, 8313–8322 (2011)

Ionic Melts in Nuclear Power

S. Katyshev and L. Teslyuk

1 Introduction

At present, there is a growing interest in studying various ways of organizing a closed nuclear fuel cycle and developing new types of nuclear power reactors and alternative technologies at all stages of nuclear fuel manufacturing and reprocessing. Molten salt reactors with fused halides with gaseous or lead coolants, various compositions of the fuel, and different spent fuel reprocessing technologies (electrolysis, high-temperature extraction, etc.) are considered. Finding an optimal prospective technology is a necessity. One of the possible fuel compositions for the fast-neutron reactors, capable of increased breeding of fissile materials, is a molten mixture uranium and plutonium chlorides with salts diluents (lithium, sodium, potassium, magnesium, calcium, and lead chlorides). Estimate calculations showed that chloride mixtures can be employed in the blanket region of a nuclear reactor [1]. The interest to the molten salt fuel is determined by simplicity and relatively low cost of the fuel production, possibility of continuous purification from the fission products, and adjustment of the molten salt composition during the reactor operation and organization of cooling. All the mentioned points were confirmed by the studies performed on MSRE and MSBR reactors in the USA [2].

The melt must have high radiation and chemical stability, low viscosity and vapor pressure under working conditions, as well as certain thermophysical properties. Molten salts are the most prospective media for the spent nuclear fuel reprocessing. Stability of molten salts toward ionizing radiation allows reprocessing nuclear fuels with short cooling times after removing from the reactor, thus implementing short fuel cycles [3, 4]. Studying corrosion stability of metals, steels, and alloys in these media showed that the problem of choosing construction materials capable of

S. Katyshev · L. Teslyuk (✉)
Ural Federal University, Yekaterinburg, Russia

© Springer International Publishing AG, part of Springer Nature 2018   181
S. Syngellakis, C. Brebbia (eds.), *Challenges and Solutions in the Russian Energy Sector*, Innovation and Discovery in Russian Science and Engineering,
https://doi.org/10.1007/978-3-319-75702-5_21

prolonged operation under working reactor conditions can be solved. At present, nickel-based alloys with high molybdenum content are considered as prospective materials for constructing a molten salt nuclear reactor [5]. However, making a final conclusion about the possibility of applying such materials is impossible without extensive additional studies of their performance in the melts containing not only uranium but also fission products.

Emerging new types of fuel and reactors result in noticeable changes of the requirements to the reprocessing technology. Therefore, perfecting reprocessing technologies and finding new technical solutions in this field remain relevant tasks. Choosing an optimal composition of salt mixtures, effective ways of cooling, and methods of spent nuclear fuel reprocessing is impossible without comprehensive studies of physicochemical, thermophysical, corrosion, and nuclear properties of the prospective salt compositions. All such information, at present, is very limited.

2 Methods Used in the Study of the Physicochemical and Thermophysical Properties of Molten Salts

The initial chemically pure salts were prepared according to the methods described earlier [6]. The surface of the primary crystallization of ternary systems was studied by differential thermal analysis with recording of the cooling curve. X-ray diffraction analysis was used for identification of the compounds.

The density and surface tension of the fused mixtures were measured by the method of maximum pressure in an argon gas bubble. The experimental technique for this measurement was well described in an earlier work [7].

The kinematic viscosity was determined employing the method based on measuring damped oscillations of a cylindrical crucible filled with the liquid salt and suspended from an elastic filament. The experimental procedure and the apparatus for viscosity measurement were described in a previous communication [8].

The method of a small bridge was used to obtain the data on thermal conductivity of the molten salt mixtures. This method is based on the measurement of the change in resistance caused by electrical heating of a narrow bridge of the material, which joins two larger bodies of the same material [9].

3 The Results of Study of the Physicochemical and Thermophysical Properties of Chloride Melts

This chapter presents the results of a study of the melting, density, viscosity, and surface tension of molten mixture ternary systems $NaCl-UCl_3-UCl_4$ and $KCl-UCl_3-UCl_4$ and thermal conductivity of molten binary systems $NaCl-UCl_3$, $KCl-UCl_3$, $NaCl-UCl_4$, and $KCl-UCl_4$.

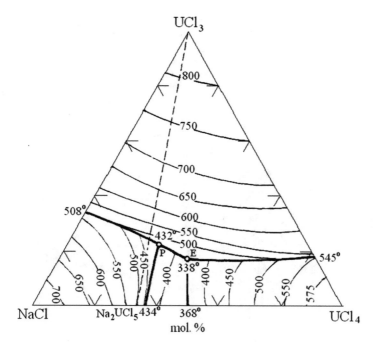

Fig. 1 Melting point diagram for the system NaCl-UCl₃-UCl₄

The surface of the primary crystallization in the system NaCl-UCl$_3$-UCl$_4$ is represented by the fields [6] NaCl, UCl$_3$, UCl$_4$, and 2NaCl • UCl$_4$ (Fig. 1). There is one peritectic point and one eutectic point for this ternary system. The eutectic in this system occurs at 338 °C, NaCl-UCl$_3$-UCl$_4$ (42.5–17.0-40.5 mol.%). Peritectic point at 432 °C corresponds to the composition (mol.%): NaCl, 49.0; UCl$_3$, 21.5; and UCl$_4$, 29.5.

In the system KCl-UCl$_3$-UCl$_4$ [7], liquidus surface is represented by five fields of crystallization: KCl, 2KCl • UCl$_3$, UCl$_3$, UCl$_4$, and 2KCl • UCl$_4$ (Fig. 2). Two quasi-binary sections 2KCl • UCl$_3$-UCl$_3$ and 2KCl • UCl$_4$-UCl$_4$ divide the system KCl-UCl$_3$-UCl$_4$ into three secondary subsystems which have one eutectic point in each of them. One (E$_1$) with a composition of 74.5% KCl + 7.5% UCl$_3$ + 18.0% UCl$_4$ has a melting point of Tm – 795 K (522 °C). Composition of second eutectic point (E$_2$) is 57.0% KCl + 17.0% UCl$_3$ + 26.0% UCl$_4$, which has a melting point of Tm – 791 K (518 °C). Eutectic point (E$_3$) at 587 K (314 °C) corresponds to the composition: 49.5% KCL + 5.5% UCl$_3$ + 45.0% UCl$_4$.

The concentration dependencies obtained for the density in the melts NaCl-UCl$_3$-UCl$_4$ and KCl-UCl$_3$-UCl$_4$ [10, 11] for 1050 K are presented in Fig. 3. The density of these mixtures increases with increasing UCl$_3$ concentration.

Figure 4 shows the lines of equal dynamic viscosity of molten mixtures NaCl-UCl$_3$-UCl$_4$ [12] and KCl-UCl$_3$-UCl$_4$ [13]. The dynamic viscosity was calculated using the results of the investigation of the density and kinematic viscosity.

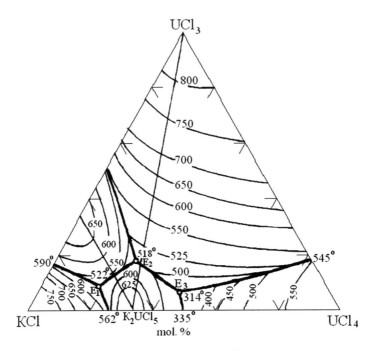

Fig. 2 Melting point diagram for the system KCl-UCl$_3$-UCl$_4$

The complicated concentration dependence of the viscosity is determined by the content of uranium chloride. The dynamic viscosity of ternary systems under study increases with increasing UCl$_3$ concentration in the mixtures.

The concentration dependences obtained for the surface tension in the melts NaCl-UCl$_3$-UCl$_4$ and KCl-UCl$_3$-UCl$_4$ [10, 11] at 1050 K are presented in Fig. 5.

The surface tension of ternary systems NaCl-UCl$_3$-UCl$_4$ and KCl-UCl$_3$-UCl$_4$ decreases with increasing UCl$_4$ concentration in the mixtures. More intense changes in the surface tension is observed at concentration up to 60 mol.% UCl$_4$. Uranium tetrachloride is as a surfactant in the mixtures under study.

The data on the thermal conductivity of molten NaCl-UCl$_3$, KCl-UCl$_3$, NaCl-UCl$_4$, and KCl-UCl$_4$ binary mixtures [14, 15] are represented as the concentration dependency (Fig. 6).

The thermal conductivity monotonically decreases with increasing UCl$_3$ concentration in the mixtures and is deflected toward smaller values (Fig. 6a). In molten mixtures of uranium tetrachloride with sodium chloride and potassium chloride, the thermal conductivity isotherms have deviation from linearity in the direction of higher values (Fig. 6b).

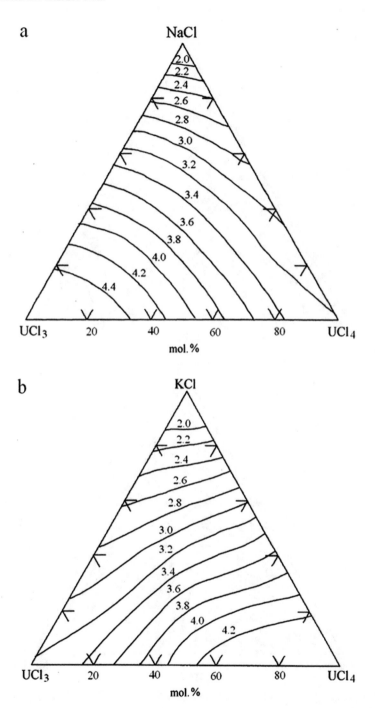

Fig. 3 Isodenses of the ternary melt of the mixtures at 1050 K, g/cm³: (**a**) NaCl-UCl₃-UCl₄; (**b**) KCl-UCl₃-UCl₄

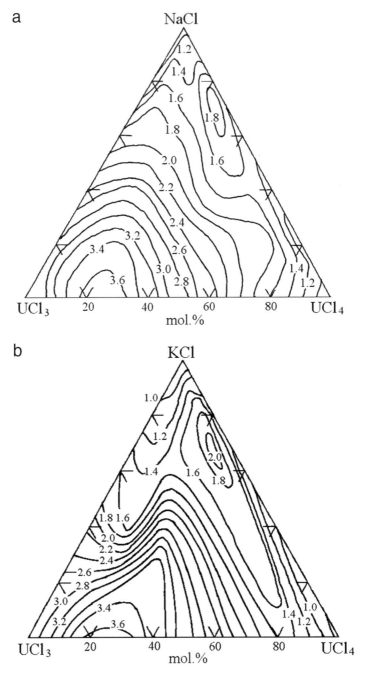

Fig. 4 Dynamic viscosity of the molten mixtures at 1130 K, mPa·s: (**a**) NaCl-UCl$_3$-UCl$_4$; (**b**) KCl-UCl$_3$-UCl$_4$

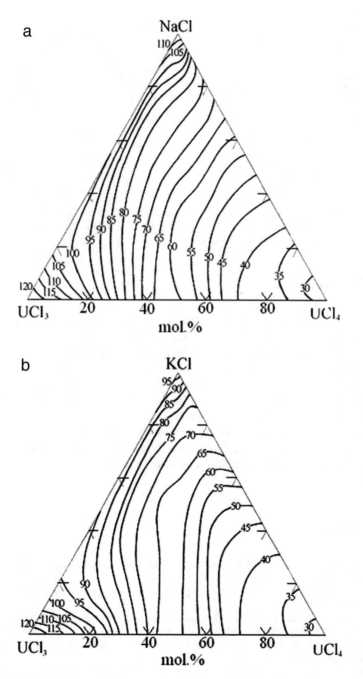

Fig. 5 Lines of constant surface tension of the melts at 1050 K, mJ/m^2: (**a**) NaCl-UCl$_3$-UCl$_4$; (**b**) KCl-UCl$_3$-UCl$_4$

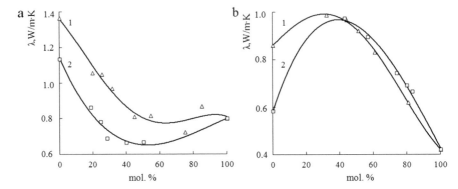

Fig. 6 Thermal conductivity (λ) isotherms of molten mixtures of: (**a**) uranium trichloride at 1123 K (1, with sodium chloride; 2, with potassium chloride); (**b**) uranium tetrachloride at 873 K (1, with sodium chloride; 2, with potassium chloride)

4 Conclusion

The presented data on the phase diagrams, viscosity, surface tension, and heat conductivity of the binary and ternary uranium containing mixtures with sodium and potassium chlorides represent part of systematic studies in this field. Magnesium, calcium, and lead chlorides can also be used as the solvent salts because they fully satisfy the nuclear-physical requirements for the fast neutron reactors. For the systems studied already, a number of low melting mixtures with high uranium chlorides content have already been found.

Studying temperature dependencies of density, surface tension, and viscosity of these salt mixtures showed that the melts containing uranium chlorides have sufficiently high density and low surface tension and viscosity.

References

1. Vasin, B.D., Katyshev, S.F., Raspopin, S.P.: Strategy of scientific and technological progress of nuclear energy. Equip. Process Indus. **7**, 11–13 (2000)
2. McNeese, L.E. (ed.): Program Plan for Development of Molten-Salt Breeder Reactors. ORNL, Oak Ridge (1974)
3. Bokavchuk, A.V., Katyshev, S.F.: Uranium chloride melts is as nuclear fuel and the environment for regeneration. In: Kharitonov, V.V. (ed.) Proceedings of the 10th International Conference on Aurorapolaris 2007. Nuclear Future: Safety, Economy and Right, pp. 349–352. MEPHI, Moscow (2007)
4. Katysheva, A.S., Katyshev, S.F.: Use of ionic melts in nuclear power. In: Krasnobaev, A.S. (ed.) Proceedings of the 11th International Conference on Aurorapolaris 2008. Nuclear Future: Technology, Safety and Environment, pp. 198–200 (2008)
5. Ignatiev, V.V., Feynberg, O.S., Zagnitko, A.V.: Molten-salt reactors: New possibilities, problems and solutions. Atom. Energy. **112**(3), 135–143 (2012)

6. Desyatnic, V.N., Mel'nikov, Y.T., Raspopin, S.P., Sushko, V.I.: Ternary systems containing chlorides of sodium, potassium, calcium, tri- and tetrachloride of uranium. Atom. Energy. **31** (6), 631–633 (1971)

7. Desyatnic, V.N., Katyshev, S.F., Raspopin, S.P.: Physicochemical properties of melt comprising mixtures of uranium tetrachlorides with the chlorides of alkali metals. Atom. Energy. **42**(2), 108–112 (1977)

8. Vohmyakov, A.N., Desyatnic, V.N., Katyshev, S.F.: A study of certain physicochemical properties of melts $CaCl_2$-CaF_2-CaO. Physicochem. Stud. Metallur. Processes Collect. Papers. **2**, 70–75 (1974)

9. Bystraj, G.P., Desyatnik, V.N.: Method of a small bridge for determination of the coefficients of the thermal conductivity of molten salts. Phys. Chem. Electrochem. Molten Salts. **220**, 56 (1973)

10. Katyshev, S.F., Desyatnik, V.N.: Bulk and surface properties of melts $NaCl$-UCl_3-UCl_4 system. Russ. J. Phys. Chem. **54**(6), 1606–1610 (1980)

11. Katyshev, S.F., Desyatnik, V.N.: Density and surface tension of melts KCl-UCl_3-UCl_4 system. Russ. J. Phys. Chem. **55**(11), 2888–2892 (1981)

12. Katyshev, S.F., Chervinskii, Y.F., Desyatnik, V.N.: Viscosity of uranium tri- and tetrachloride melts with sodium chloride. Russ. J. Phys. Chem. **57**(11), 1256–1257 (1983)

13. Katyshev, S.F., Chervinskii, Y.F., Desyatnik, V.N.: Density and viscosity of molten mixtures of uranium chlorides and potassium chloride. Atom. Energy. **53**(2), 108–109 (1982)

14. Bystraj, G.P., Desyatnik, V.N., Zlokazov, V.A.: Thermal conductivity of molten uranium tetrachloride mixed with sodium chloride and potassium chloride. Atom. Energy. **36**(6), 517–518 (1974)

15. Bystraj, G.P., Desyatnik, V.N.: Thermal conductivity of the molten mixtures of sodium chloride and potassium chloride. Atom. Mol. Phys. **44**, 125–127 (1976)

Part V
Environmental Aspects of Using Energy

A Methodical Approach to Assess the Efficiency of Renewable Energy Projects with Regard to Environmental Component

I. Rukavishnikova and A. Rumyantseva

1 Introduction

Power generation from renewable and inexhaustible sources is one of the most prospective ways of power industry development. Many experts talk about "energy transition" [1], about technologically developed economies entering the age of renewable energy. A share of renewable energy in global energy generation might be the indicator of innovative development of a country or a separate region.

Compared with most industrially developed countries, the scale of renewable energy usage and its growth speed in Russia is rather small – its share in electric account of the country, excluding major HPS, is about 0.7% of the total power generation [2]. According to RF government decree "Concerning basic government policy directions in energy efficiency boosting based on the use of renewable sources of energy until the year of 2020" [3], by 2020 this share must rise to 4.5%, but experts are pessimistic about the possibility of reaching this goal.

However, the technical potential of renewable energy in Russia is very high and is about 3000 mln toe [4]. According to research [2], the economic potential of renewable sources of energy is valued at 190 mln toe per year, which is more than 30% of the whole initial supply of energy resources [4].

The purpose of the given research was to detect the reasons preventing Russia, and particularly the Sverdlovsk region, from intensive development of renewable energy and the objectivation of the most promising ways of renewable project implementation in the Sverdlovsk region.

I. Rukavishnikova (✉) · A. Rumyantseva
Ural Federal University, Yekaterinburg, Russia

© Springer International Publishing AG, part of Springer Nature 2018
S. Syngellakis, C. Brebbia (eds.), *Challenges and Solutions in the Russian Energy Sector*, Innovation and Discovery in Russian Science and Engineering,
https://doi.org/10.1007/978-3-319-75702-5_22

2 The Prospects of Renewable Energy Development in the Sverdlovsk Region

The Sverdlovsk region has a significant technical and economic potential for renewables' implementation [5, 6].

The region has a large headwater and reservoir (more than 400) system. Therefore it has good options for small hydropower development.

There are areas with a good wind probability in the North and mountain part of the region's territory that allow the use of wind turbine generators (WTG). Experts estimate WTG feasible total capacity at 200 megawatts.

The income of solar radiation in summer daytime is about 400–650 watt per sq. m. It allows the use of solar power in private houses and in supplying electricity to essential customers with photovoltaic systems.

It is worth noting that the economic potential of wind and solar power projects in the region is much lower than the technical potential [6]. Primarily, it is due to the region's geographical and climate specific characteristics and the risks connected to them.

Bioenergy might be considered as one of the most prospective branches of renewables in the region. There is a tendency to use timber processing waste energy in the timber industry instead of gas fuel.

The region is one of the leading regions of Russia in livestock breeding development. There are several major poultry and livestock production units in the region, and their processing is connected to making substantial volumes of biological waste. Existing technologies of biogas production from livestock and poultry waste allow the production of more than 700 toe a day.

It is worth noting that the problem of livestock waste handling is rather topical, both in the Sverdlovsk region and in the whole country. Nowadays, in the best-case scenario, it is being burnt, but usually it is stocked in special tanks in fields or simply in the farms' backyards. This way of disposal leads to significant ground, water, and air pollution and to alienation of large areas for waste stocks. Using waste in bioenergy projects seems to be the most civilized decision.

In 2011, the strategy of fuel-energy complex development [5] in the Sverdlovsk region by 2020 was passed. The main goals of this document are to create a program of renewable energy development in the region; setting the target values of the program in building up and using renewables by types, quantity, capacity, and productivity; reaching these targets before 2020; and large-scale assimilation of all the renewables' potential types based on economic viability.

3 The Factors of Making Decisions About the Implementation of Renewable Energy Projects

As part of the study, we interviewed experts in renewable energy – scientists of Ural Federal University, whose scientific interest area is renewables and energy-saving, regional and municipal administration officials, and practical specialists who work in the implementation of renewable energy projects. In total we interviewed 20 people.

According to most specialists' opinion, the most critical problems that slow down development of renewable energy in Russia are:

1. The absence of renewable sources of energy law
2. The imperfection of existing mechanisms of state economic support for renewable energy projects
3. A low level of regional officials' commitment
4. A low level of awareness and economical motivation of persons and small- and medium businesses in renewable projects' implementation

The present situation might be explained by some objective factors: the country's geographical position, which gives us low insolation and small speed of winds in more than 65% of the country's areas, large reserves of hydrocarbons, a developed network and capacity of fuel energy and large hydropower plants, and fast growth and unprecedented innovative potential of nuclear energy. Alternative energy projects are unlikely to become an economical competitor to conventional energy, although some of them definitely have potential for implementation.

Figure 1 gives results of expert assessment for the importance of factors of making decisions about implementation of renewable energy projects.

The left bar of each factor is its important assessment by scientists, specialists in renewable energy and energy-saving; right bar of each factor is its important assessment by experts who implement renewable projects. The factor with the lowest grade given by both expert categories can be considered as top priority in making decisions about the implementation of a renewable project.

According to most specialists, the decisive factor of project implementation is economic efficiency of the project; next by importance is acquiring independence from a central electrical and heating supply. Potential reduction of energy charge due

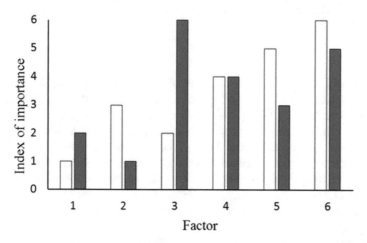

Fig. 1 Results of ranging factors of renewable implementation a priori: 1, economic efficiency of the project; 2, providing independence from central electric and heating supply; 3, pollution reduction; 4, formation of "environmentally responsible organization" image; 5, reduction of fossil fuels exhaustion speed; 6, reduction of energy charge

to usage of renewables has the lowest impact on decision-making, according to specialists (in the opinion of most specialists, that is not the most important criteria of renewable project's economic effect in Russia, unlike most of energy-saving projects). Strange though it may seem, environmental factors, such as pollution reduction, reduction of fossil fuels exhaustion, and formation of "environmentally responsible organization" image, happened to be of small importance, too. We think that this stems from imperfection of existing state and regional economic mechanisms of stimulation for environmental activities, one of which, in our opinion, is implementation of renewable projects. This problem can be solved by adequate economic introduction of environmental data in project efficiency.

4 Methodical Approaches to the Assessment of Renewable Energy Projects' Efficiency

We could not find any specific approaches to the assessment of renewable energy projects' efficiency in scientific and methodical literature, but it seems obvious that these projects are energy-saving and environmental by nature. So, we think that it suffices to assess them with approaches developed for energy-saving and environmental projects in addition to traditional methods of investment projects' assessment.

According to some authors [7, 8] the assessment of energy-saving projects' efficiency requires the development of specialized methodical approaches. Semionov [7] offers to take the cost per unit of saved mineral and energy resources as a defining criterion of energy-saving projects' efficiency. The ratio of total expenditures for a given project implementation ($\sum C_{pr}$) to volumes of saved resources ($\sum V_{res}$) in a period of project existence is cost per unit (C_{unit}) of saved energy (Eq. 1):

$$C_{unit} = \sum C_{pr} / \sum V_{res} \qquad (1)$$

While choosing the project for funding, primary favor should be given to enterprises with lower costs [7]; then, while implementing these projects, one can start less economically viable projects. If costs of project implementation exceed maximum profit, these projects are technically possible, but ineffective economically. Their implementation can be justified only by an extreme level of their environmental importance. In the authors' opinion [7], the criterion of environmental efficiency of energy-efficient project defines its potential profit, which appears in reduction of negative influence on environment while reducing usage of energy resources or water. Quantitative evaluation of the given criterion depends on calculation methods employed.

Telyashova and Kosmatov [8] think that efficiency assessment of energy-saving projects must be complex. They outline these groups of factors that influence complex efficiency of energy-saving projects: economical, technological, environmental, and social. The variety of taken factors and efforts of its systematization is surely worth of attention, but the ability to economically evaluate most of the factors and determination of their intake to economic result of energy-saving project realization causes some doubts.

The usage of renewable sources of energy, as well as most other projects in the area of energy-saving, can be related to environmental projects. This means that we can use approaches formed for assessment of environmental activities' efficiency in assessing renewable energy projects.

One of the main criteria of choosing an environmental project is its high environmental-economic efficiency.

Calculation of environmental-economic efficiency allows us to determine the effect reached by energy-saving technologies not only with the help of funding rationality determination but also taking into account economic evaluation of technologies' effect on resource-saving, including saving the quality of environmental components.

One of the main economically assessed results of environmental projects is prevented (solar energy, wind energy) or significantly lower (biofuel technologies) environmental damage to atmosphere connected to pollution. Some technologies (biofuel, waste usage) also prevent damage to land resources connected to toxic waste and land degradation.

The traditional environmental result of the projects is the reduction of emission charge.

Unfortunately, these charge reductions are usually too low to have significant effect on project efficiency; and environmental damage prevention mostly brings the implementer only moral satisfaction.

One of the goals of this research is a complex assessment of a renewable project planned for implementation. The purpose of assessment is choosing the project with high economic and environmental efficiency.

5 Efficiency Assessment of Bioenergy Project Based on Livestock Waste Usage

In order to develop an approach to the efficiency assessment of a bioenergy project, we used data from the project of biogas plant for recycling of livestock organic waste from Sverdlovsk region's farms, which was developed for business plan level by LLC "AlBIOt" but was not implemented yet due to some reasons [9].

The point of this investment project is creating a new plant for recycling livestock organic waste from farms – manure mixed with straw and slaughterhouse waste. Basic technological equipment is biogas unit (BGU), which should be manufactured and installed by LLC "Dzeta-service."

In the course of assessment, we defined sources of income [9]: (1) selling liquefied biogas (methane) and (2) selling dry and liquid biofertilizers depending on the season. Also we considered abilities of extra profit from BGU exploitation: (1) electricity and heating generation (with extra modules installed), (2) using methane as automobile fuel (with extra modules installed), and (3) environmental profit (while recycling).

Expenditures of the project were all the operating costs, including salaries, building maintenance, office functioning costs, etc. The total amount of funds required for project implementation is about 180 mln rub. excluding cash for investment period [9].

We analyzed the market in order to certificate costs and profits of the project. The most advanced operations in agricultural sector of Sverdlovsk region economy are food animal farming (cattle population reached 187,700 heads, pig population reached 285,200 heads in 2014; 259,400 tons of cattle and poultry was produced in all farm categories in 2014), grain and bean growing, potato and vegetable farming, and technical crop cultivation. The planted acreage in 2014 reached about 900,000 ha [10].

It is obvious that efficient crop farming is impossible without fertilizers, and livestock farming, from environmental point of view, is impossible without disposal of livestock (cattle, pigs, poultry) waste.

Mineral fertilizer consumption in Sverdlovsk region tends to be lower due to the decrease of total planted acreage in the region. Negative tendency in organic fertilizers usage in 2008 started to change to positive due to limitlessness of their appliance and insertion on the basis of availability. The organic fertilizers market is one of the most dynamic and prospective.

The unit suggested for the investment project can produce 23,600 tons of organic fertilizers per year, which is about 1.5% of fertilizers consumed in the Sverdlovsk region in a year.

The suggested unit can recycle 50,400 tons of waste per year, which is 1.8% of total waste required recycling (about 2.5 mln tons of waste was produced in the Sverdlovsk region in 2014 [10]). This means that for recycling of all waste produced in the region, we need about 50 biogas units. It is worth noting that replication of tried and true renewable projects is widely spread in some countries – leaders in renewable energy. It also helps to lower costs per energy unit. So, the researched project can be assessed as quite promising.

Calculated integral financial data (Table 1) shows the high economic efficiency of the project.

Project efficiency connected to energy-saving can be pre-estimated using the data of plant capacity forecasts [9]. The unit should produce 2.78 mln m^3 of biogas per year. Calorific equivalent of biogas is close to natural gas fuel value. So, natural energy resource (natural gas) savings resulting from project implementation can reach 2.78 mln m^3 per year in physical terms and, according to current gas wholesale tariffs for 2015, 9060 mln rub. per year in money terms. During the life of the project, it will reach 49,830 mln rub. Costs per unit of saved energy resource in money terms, defined in Eq. (1), are 3.61 rub. per rub. We can conclude that the project is economically viable not only due to its energy-saving but also due to the complexity of achieved results. The project has reserves for upgrade of energy-saving performance. Along with livestock waste, we can use poultry manure and crop farming waste that significantly increase biogas production (Table 2).

Table 1 The project's integrated financial data

Project indicators	Indicator value
1. Discount rate (%)	7
2. Life of the project, month	66
3. Simple payback period, month	38
4. Discounted payback period, month	40
5. Net present value, thou. rub.	221,190
6. Internal rate of return (%)	42.3
7. Profitability index	2.1

Table 2 Biogas production of the unit [9]

Substrate	Biogas production, m^3/ton
Cattle manure (natural 85–88% hum.)	60
Pig manure (natural 85% hum.)	65
Corn silage	180–220
Bird manure	80–140
Fruit pulp (80% hum.)	70
Beet pulp (77% hum.)	100

The main environmental result of the project is recycling of accumulated and newly produced waste. As a result, accumulated environmental damage is liquidated, and land resource damage, connected to waste dumping and land degradation, is prevented. Nowadays defined damage value has a significant effect on economic results only if environmental regulators fine one's business for unauthorized livestock waste dumping. Charge for waste dumping (even in case of over-limit dumping) also does not have significant effect on financial results. So we can conclude that approaches to calculation of environmental component of project efficiency should be developed in such a way that environmental result would bring in a real contribution to the project's economic efficiency. In a case of bioenergy projects based on waste usage, this can be achieved by stricter control of waste dumping, increased charges for over-limit dumping, and fines for unauthorized waste dumping. Environmental efficiency nowadays does not have a significant impact on project financial data and can be seen as a defining factor of implementation only in case of acquiring financial aid from municipal, regional, or federal government.

6 Conclusion

The key criterion in the assessment of renewable project efficiency and, consequently, the key factor of making decisions on project implementation is the economic efficiency of the project. Economic results from energy-saving and environmental efficiency of the project can be used as the additional selection factors for projects with similar economic efficiencies.

Acknowledgment The work was supported by Act 211 Government of the Russian Federation, contract № 02.A03.21.0006.

References

1. Smil, V.: Energy Transitions. History, Requirements, Prospects. Praeger, Santa Barbara (2010)
2. Bezrukih, P.P., Karabanov, S.M.: Role of renewable energy in Russian industry modernization. Business Pride Russia. **5-1**(48), 40–42 (2015)
3. Decree of the Government of RF on January 08 (2009)
4. Renewable energy in Russia. From possibilities to reality. OECD/IEA (2004)
5. Strategy of fuel-energy complex development in Sverdlovsk region by 2020. http://pandiaweb.ru/text/78/502/98534.php
6. Low-carbon future for Sverdlovsk region. Urals environmental council, Yekaterinburg (2012). http://rusecounion.ru/sites/default/files/UES_48.pdf
7. Semionov, V.N.: Analysis and assessment of economical and environmental efficiency of energy-saving project sin the system of essential service sofa municipal unit. Bull. Kazan State Univ. Arch. Construct. **1**(13), 380–384 (2010)
8. Telyashova, V.S., Kosmatov, E.M.: Methods of Efficiency Assessment and Stimulation of Innovative Energy-Saving Technologies in Power Generation and Transportation. Saint-Petersburg State Polytechnic University, Saint-Petersburg (2010)
9. Business-plan for investment project of BGU Alapaevsk. LLC AlBIOt, Yekaterinburg (2010)
10. The official website of Ministry of Agriculture and Food Supplies of Sverdlovsk Region. http://mcxso.midural.ru/paper/show/id/105

A Concept of Transition to a Technological Regulation System for the Power Industry

M. Berezyuk and A. Rumyantseva

1 Introduction

In accordance with the Federal Law № 219-FZ "On Amendments to the Federal Law 'On Environmental Protection' and Certain Legislative Acts of the Russian Federation," a gradual transition to the technological regulation system, based on the best available technologies (BAT), using the European Union experience and specifics of the domestic economy began in Russia since January 1, 2015 [1]. The goal of this research is the use of a systematic approach to the problem solution and the development of a mechanism to select the best options for the introduction of new technologies based on BAT.

2 Transition to Technological Regulation System

Power industry of the Russian Federation is among the three leaders in terms of the exerted influence volume. Major energy sector companies are classified as units with significant impact on the environment. As a part of the energy strategy of Russia for the period until 2030 (approved by the Government Decree № 17–15-r of November 13, 2009), the target figures to reduce the volume of the industry's negative impact on the environment were developed [2].

To reach the set figures, radical changes in the monitoring approaches to the companies' impact on the environment and stimulation of this impact reduction are needed. According to the authors, the transition to a regulation system, based on

M. Berezyuk (✉) · A. Rumyantseva
Department of Environmental Economics, Ural Federal University, Yekaterinburg, Russia
e-mail: m.v.berezyuk@urfu.ru

© Springer International Publishing AG, part of Springer Nature 2018
S. Syngellakis, C. Brebbia (eds.), *Challenges and Solutions in the Russian Energy Sector*, Innovation and Discovery in Russian Science and Engineering,
https://doi.org/10.1007/978-3-319-75702-5_23

Table 1 Stages of transition to the technological regulation system

Period	Work-stage contents
2015–2017	The division of the companies into four categories according to the degree of their negative impact on the environment State registration of working companies and category conferment The development of BREF reference documents
Since January 2015	State regulation measures in the field of environmental protection will apply only to the pollutants included in the list established by the government of the Russian Federation
2019–2022	The owners of category I companies, with the contribution to the total pollutant emissions not less than 60%, are required to apply for an integrated environmental permit (IEP)
Since January 2019	The coming into force of the legal requirements for the inclusion into IEP of the mandatory program of eco-efficiency increase and environmental protection action plans for the facilities of category II
Until January 2025	The rest of the economic entities will have to get IEP

BAT, can make a significant contribution to the achievement of these goals. The foundation of these changes relies on the principles and provisions of environmental regulation, approved by the European Union Directive on Industrial Emissions 2010/75/UE (IED) (integrated pollution prevention and control, IPPC) [3], instead of the previously used EU Council Directive № 96/61/EU of 24.09.1996 "On Integrated Pollution Prevention and Control (IPPC)" [4]. According to the current IED, only big thermal stations with the thermal energy exceeding 50 MW are taken into account, thus excluding the other types of stations such as nuclear or hydro-electric power stations.

Certainly, the Russian Federation, in spite of certain achievements, stands at the beginning of transition to technological regulation; that is why the experience of the EU is carefully studied and adaptation of the EU regulatory framework to the legal field and the economic and technological characteristics of the Russian Federation is carried out. The planned stages of the transition to technological regulation system are shown in Table 1.

As mentioned above, the direct use of European BREFs [5, 6] by the Russian companies is hardly possible due to the existing differences in characteristics of all types of resources, peculiarities of raw materials, the availability of different energy types, natural conditions, environmental characteristics of the areas, and production technologies. That is why a science-based methodological support should be developed for transition of the Russian energy sector to BAT. An algorithm, consisting of several stages, is proposed. The algorithm is original and rather subjective, but it helps to choose the best technologies in accordance with a systematic approach, taking into account economic and environmental components of this process.

3   Application of BAT Selection Algorithm by Energy-Producing Companies

To determine BAT, the most effective technology (technical measures, administrative decisions) in terms of achieving a high general level of environmental protection at the energy-producing companies should be chosen.

Stage 1. Current situation of the companies' analysis. It is necessary to estimate the current situation of the companies in terms of their impact on the environment. First of all, environmental pollution at the power companies depends on the technology used at the power station and fuel type that was used to generate energy. On the basis of the company impact volume on the environmental components, its risk category should be determined from an environmental point of view. The volume of the impact on the environmental components also should be defined, so it would be possible to determine whether the impact of the given company is subject to the BAT regulation [5].

Stage 2. Determine the sphere of application for alternative technologies. At this stage it is necessary to consider all technologies that can be used to reduce the impact on the environment [5, 6]: (1) technological solutions, (2) raw material selection, (3) production processes control, (4) organizational arrangements, (5) "nontechnical" events, and (6) end-of-pipe technology. It is likely that at this stage the level of impact on the environment and the possible effects of the technologies introduction will become evident. It will be possible to choose the most appropriate technology [7, 8].

Stage 3. Alternative technologies analysis. At this stage, data on pollutant emissions (discharges, wastes production, and consumption) as a result of considered technologies application should be analyzed and summarized, as well as the data on used resources. During this stage of realization, relevant input and output parameters should be submitted in the form of a list (with quantitative indicators) of the considered technologies. This list should include produced discharges, emissions, wastes, other impacts, and consumed materials (water, coal, gas) [8].

Stage 4. Evaluation of all types of environmental impacts. Comparison of the different pollutants is conducted for each of the considered alternative technologies according to seven priority environmental problems: (1) toxicity for people, (2) the toxicity of water bodies, (3) global warming, (4) acid rain formation, (5) eutrophication, (6) ozone layer depletion, and (7) probability of tropospheric ozone formation. A reference publication [6] describes the integrated assessment methodology of the technological impact on the environment, which can be used to compare alternatives considered as BAT.

Stage 5. Description of the environmental problems assessment approach. Three possible approaches are proposed to evaluate options and get the results on the basis of the fulfilled assessments. Each of the approaches can be used independently or together. The first approach is the most simplified; it compares the previously considered and calculated impacts for each of the seven environmental problems.

The second approach is more complex and allows one to compare contributions made by the considered technology for each of the seven environmental issues, with all-European indexes. The third approach allows one to compare considered pollutants separately with the data of the European Pollutant Release and Transfer Register [5, 6].

Stage 6. Analysis of the additional information for alternative technologies description. At this stage all available additional information to clarify the description of the technology is gathered: technical and economic service time of the equipment and service data. Detailed characteristics will be used to collect and analyze data on costs. Together with the level of impact on the environment, it is necessary to assess the degree of exploitation reliability of considered production systems for the local area [8].

After reviewing and ranking of the possible options in terms of environmental performance, the variant with the least impact on the environment is considered to be the best, but only if this option is available from the economic point of view. Therefore, after the integrated assessment of impacts on the environment, it is necessary to evaluate and compare the costs of the considered alternative technologies implementation [7].

Stage 7. Data collection on the costs of the technologies introduction. All factors that may affect the data accuracy should be taken into account. Certainly, this may affect the evaluation results and the final decision on BAT selection. The main sources of the cost data are technology or equipment producers (suppliers), consultants and research groups, energy sector development plans, authorities, published information (papers, reports of engineers, sites of the power complex companies, data from conferences), and costs assessment of the comparable projects in other industries [8].

Stage 8. Determination of the cost structure of technologies implementation. The main task of this stage is to determine which elements of the costs should be included or excluded from the assessment. This stage helps to understand the cost structure and items to which costs are allocated. It is possible to distinguish the following groups of expenses: (1) investment, (2) annual expenditures for operation and maintenance (operating costs), and (3) costs to be considered separately.

All costs are to be assessed relatively to the basic version (the existing situation). The basic version is to be set according to the BAT assessment methodology, while the alternative versions are considered relatively to the basic version. Expenditures for all versions are to be shown for construction of new power plants.

Stage 9. Revenue assessment. It is necessary to collect information on additional income that may arise with the BAT implementation: (1) sales revenues, (2) avoided costs, and (3) following benefits.

Stage 10. Processing and presentation of information on costs and revenues. The collected information on costs and revenues should be processed in such a way that it would be possible to objectively compare alternatives. It is possible to formulate the most significant moments of information processing and presentation: (1) data calculation and presentation in the form of annual costs and revenues, (2) stating cash flows in a single currency (rubles for the Russian Federation), (3) adjustments for inflation effects, (4) use of the discounted cash flow method, and (5) valid approach to the discounting rate determination [8].

Stage 11. Evaluation and comparison of alternative technologies. After the benefit for the environment and the economic costs on the implementation have been established for the alternative technologies, it is necessary to fulfill a comparative analysis and to determine which technology meets the BAT criteria [6, 7]. There are several major ways to determine the cost-effectiveness with reference to the Russian energy sector companies.

Economic effectiveness analysis. In the context of ecological policy, this method means the achievement of the highest ecological results for every ruble invested into the environmental measures.

The most obvious way to compare the costs of the activity realization and the gained benefits consists in presenting them in the pecuniary form and comparing them by the analysis of costs and gains. If the comparison shows that the benefits exceed the costs, then the activity is worth investing into. Though such an analysis requires the large amount of data, and some benefits are difficult to present in the pecuniary form.

The cost efficiency analysis is simpler than costs and gains analysis as in this case the ecological benefits are assessed quantitatively and not in money terms. The economical effectiveness is defined by the following: economical effectiveness = annual costs/emissions (discharges) decrease.

The allocation of the costs between the polluting emissions (discharges). This brings the additional information regarding the methods of the cost allocation between the polluting substances, the ingress of which into the environment should be prevented or decreased. If the costs related to the environmental technologies were allocated between the polluting substances, then the method of their proportional distribution should be described. There are two possible approaches to the cost allocation:

1. The costs of the technology (equipment) may be fully allocated to the same problem of the environment pollution, for which this activity was initially intended. Then the decrease of the other pollutant discharges will be considered the additional benefit for the environment as it did not require the additional costs.
2. The scheme of the cost apportionment may be created to allocate the costs between the pollutants, the impact of which on the environment provokes the concerns.

The comparison of the technologies introduction costs and environmental benefits. The technology will be considered effective if the environmental benefits will exceed its implementation costs.

The cost-effectiveness of the technology introduction equals the sum of the discounted income over the sum of the discounted costs. But in this case there are certain difficulties with the justification of the discount rate and the amounts of revenues and expenses for an extended period of time. But even considering these difficulties, according to the authors, the method is the most preferable and reflects reality most accurately.

The technology identified as BAT should be developed on such a scale which allows its implementation in the relevant industrial sector, under economically and technically valid conditions. The assessment of the economic efficiency is only necessary with BAT determination at the industry level, when the proposed technologies lead to the fundamental changes in the energy sector.

4 Evaluation of the Technologies Economic Viability in the Energy Sector

A reference publication [6] and the national standard of the Russian Federation [9] provide general recommendations on the assessment of the technologies economic cost-effectiveness in various industry sectors. In this chapter the authors focus on the peculiarity of such assessment in the energy sector. In the authors' opinion, this approach will allow a better understanding of the existing differences in the types of resources, peculiarities of raw materials, the availability of different energy types, natural conditions, environmental and production characteristics of the areas, and a better disclosure of the BAT transition concept.

It is assumed that for the consideration of the economic cost-effective assessment at the industry level, the most significant problems are: (1) industry structure, (2) market structure, (3) the ability of quick recovery ("elasticity"), and (4) implementation rate [6, 9] (Fig. 1).

The structure of the (energy) industry describes the socioeconomic characteristics of the considered industry and the technical characteristics of the industry

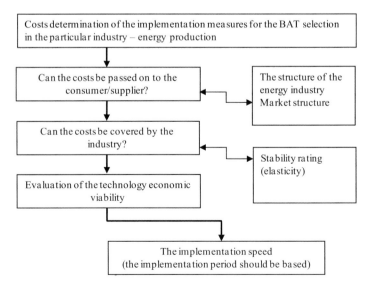

Fig. 1 The evaluation of the BAT implementation economic cost-effectiveness in the energy sector

companies. These characteristics allow a better understanding of the industry structure and how easily can the BAT implementation pass. While describing the structure of the industry, it is logical to consider the following questions: (1) the size and number of the companies in the industry (large companies are common for the energy industry). (2) Companies' (units) specifications. The existing infrastructure of the company will have some impact on the BAT type and can also affect the implementation costs of this technology. (3) Equipment service life. The equipment with long service life is usually used in the energy sector. It is a determining factor in the investment cycle. (4) Barriers for market entry or exit. In cases when there are barriers for the new players' entry into the market (e.g., high prices for equipment or licenses), the problem of technical barriers should be considered separately.

The market structure can affect the ability of the company in terms of the transfer and assignment of the "environmental" costs of the BAT implementation on to the consumers or suppliers. There is a range of factors characterizing the market structure for the energy sector. Many of these factors are associated with a qualitative assessment, which makes it difficult to determine their effect on BAT; but among them the most important factors can be marked: the market size, the price elasticity, and the competition between the products. The analysis of the market structure facilitates the significant risk identification and allows industry to consider the impact of these risks (if any) for the BAT definition.

The industry stability (elasticity). The elasticity reflects the ability of the industry to cover the increasing costs of the BAT implementation, while maintaining its profitability in the short, medium, and long term. Any increase in costs associated with the BAT implementation should be covered by the industry or passed onto the consumer. The elasticity describes the ability of the industry to cover these costs. To describe the elasticity of the industry, it is useful to consider the long-term trends (5–10 years) to ensure that short-term fluctuations do not affect the BAT determination.

The implementation speed. If after assessing the industry structure, market structure, and stability of the industry certain doubts about the idea of BAT remain, the technology implementation speed can be estimated as the implementation time can be critical for the energy sector. The implementation speed normally is not an issue for the new facilities (as opposed to active) because it is expected that new companies are ready to use environmentally safe technologies and pollution control equipment. Therefore, the evaluation process should distinguish new and existing enterprises. To determine the BAT implementation speed, it is useful to consider the following time scales: short term (several weeks or months), medium term (from a few months to a year), and long term (several years).

5 Conclusions

The economic cost-effectiveness is an integral part of the best available technologies concept. But its in-depth assessment should be carried out only in cases when there are clear differences regarding which BAT can be implemented cost-effectively in the energy sector. The approaches considered by the authors are quite reliable and form a structure for the decision-making process. The concept of the transition to BAT presented in this chapter should help to clearly state the existing aspects, to justify the costs and benefits of the technologies implementation and ensure the further energy companies development.

References

1. Federal Law № 219 of 21.07.2014. On amendments to the Federal Law "On Environmental Protection" and certain legislative Acts of the Russian Federation. http://www.rg.ru/2014/07/25/eco-dok.html
2. The Energy Strategy of Russia for the period until 2030, approved by the Government Decree № 17-15-r of November 13, 2009. http://minenergo.gov.ru/aboutminen/energostrategy
3. Directive 2010/75/UE (IED). http://ec.europa.eu/environment/industry/stationary/ied/legislation.htm
4. Directive 96/61/EC (IPPC). http://eurlex.europa.eu/LexUriServ/LexUriServ.do?uri=CELEX:31996L0061:en:html
5. Handbook on Best Available Techniques for Large Fuel Combustion Plants (transl.), [in Russian], SE "INVEL", Moscow (2009)
6. Economic aspects and impacts on different components of the environment. Reference Document on Best Available Techniques, [in Russian], Project "Harmonisation of environmental standards HES II, Russian" (2006). http://14000.ru/brefs/BREF_ECME.pdf
7. Sokornova, T., Koroleva, E., Sergienko, O., Kryazhev, A.: Economic aspects of the BAT implementation, [in Russian], Ecol. Prod. 10, 28–35; 11:44–49 (2012)
8. Berezyuk, M., Rumyantseva, A., Merzlikina, J., Makarova, D.: The development of an ecological-economic substantiation algorithm for BAT selection for companies within the power industry. WIT Trans. Ecol. Environ. 190(2), 1161–1172 (2014) WIT Press, UK
9. National standard of the Russian federation, [in Russian], Resources saving. Identification methodology. http://docs.cntd.ru/document/gost-r-54097-2010

The Stimulation of Renewable Energy Source Usage: Economic Mechanism

Andrey Boyarinov

1 Introduction

According to the long-term prediction of the socioeconomic development of the Russian Federation up to the 2030s, in the twenty-first century, energy security will become a battleground, initiating a new conflict or cooperation. Future stability and the possibilities of the country's economic growth will be defined by the success of the response to such global threats such as global climate change and a lack of energy.

Being connected to all the vital services (healthcare, food, transport, and trade), energy will be the key factor of global economy development. Energy demand will grow globally. Lack of conventional strategic resources (water, minerals, metals) will be felt stronger. The search of long-term sources of energy will take more effort. Investments in development of new energy sources can attract billions of dollars. Economic growth and labor productivity will decrease if decisions on the use of alternative fuels are not made. The development of energy technologies might have a significant effect not only on the energy industry but also on the field of the environment [1].

One of the main prospective development vectors of the fuel and energy industry branches, provided by the energy strategy of Russia for 2020, is the transition to the way of innovative and energy-efficient development [2].

Renewable sources of energy (renewables) are energy resources of constantly existing natural processes on the planet and energy resources of waste products of plant and animal biogenesis. The main characteristic of renewables is their

A. Boyarinov (✉)
Department of Environmental Economics, Ural Federal University, Yekaterinburg, Russia
e-mail: au.boyarinov@net-ustu.ru

© Springer International Publishing AG, part of Springer Nature 2018
S. Syngellakis, C. Brebbia (eds.), *Challenges and Solutions in the Russian Energy Sector*, Innovation and Discovery in Russian Science and Engineering,
https://doi.org/10.1007/978-3-319-75702-5_24

inexhaustibility or their ability to recover potential in a short time (within the period of one people's generation lifetime).

The energy of solar radiation, winds, water flows, and biomasses and the thermal energy of the upper layers of the Earth's and ocean's lithosphere are referred to as renewables.

2 Trends and Forecasts of Renewable Energy Usage

The practicability and scales of renewable usage are primarily defined by their economic efficiency and competitiveness with traditional energy technologies. The main advantages of renewables, in comparison with fossil fuel energy sources, are the actual inexhaustibility of these resources, ubiquity of most of them, and absence of fuel costs and emissions. However, they are more capital-intensive, and their share in total energy production is not big yet (except for hydropower plants). According to most forecasts, this share will remain moderate in the coming years. At the same time, interest in the development and implementation of unconventional and renewable energy sources will increase in many countries (Table 1).

About 100 countries have special state programs of renewable implementation and indicatives of their development, approved at the state level for medium- and long-term prospective. Most countries set targets to raise renewables' contribution to the energy balance of the country at least to 15–20% by 2020 and European Union countries to 40% by 2040. Top-priority development of renewables with the growth rate of 10% per year is implemented with powerful legislative, financial, and political support from the government.

Nowadays, Russia is almost absent in the renewable energy market. The contribution of unconventional renewables (except large HPP) to energy balance of Russia is less than 1%. Recent governmental decisions order to raise renewable contribution to 2.5% by 2020 (Table 2).

That will demand creating power plants working on renewables with a total power of 20–25 GW. Unlike many other countries, there is no clear and coherent state policy formed for renewables' usage in Russia. Political declarations about

Table 1 Indicators of global renewable energy [3]

Indicators	Year			
	2009	2010	2011	2012
Investment in new projects (per year), billions $	161	227	279	244
Renewable power plant capacity (without HPP), GW	250	315	395	480
Energy production from biomass, TWh	–	313	335	350
Photovoltaic power plant capacity, GW	23	40	71	100
Solar power station capacity, GW	0.7	1.1	1.6	2.5
Wind farm capacity, GW	159	198	238	283
Number of countries using renewable energy, unit	85	109	118	138

Table 2 Forecast indicators of input capacity of renewables in Russia (MW) [4]

Types of renewables	Installed capacity	2015	2016	2017	2018	2019	2020	Total
Hydropower plant (capacity less than 25 MW)	18	26	124	124	141	159	159	751
Wind farm	100	250	250	500	750	750	1000	3600
Solar power plant	120	140	200	250	270	270	270	1520
Total	**238**	**416**	**574**	**874**	**1161**	**1179**	**1429**	**5871**
Percentage of total installed capacity, %	0.10	0.28	0.51	0.88	1.37	1.87	2.48	

renewables' importance are still not thoroughly reinforced by a set of legislative acts and regulations which stimulate renewable usage and define order of cooperation for investors and consumers. The relation toward renewables is contradictory in Russia. There are enthusiasts who claim that we need to use renewables as widely as possible at the very moment, and there are pessimists, mostly from the fuel and energy complex industry, who claim that renewables are not very promising for Russia, energy nation with large reserves of organic fuels, and that they won't be able to make a significant contribution to the energy balance of the country in the nearest future, and that is why they shouldn't be seriously developed yet [5].

3 Restrictions of the Renewable Energy Production

Intensive development of renewables in Russian Federation is constrained by a number of barriers. All of them could be divided into three groups: financial, information, and institutional barriers.

3.1 Financial Barriers

The main problem is the lack of domestic and foreign investments. Russian companies interested in the development of renewable usage have limited financial resources and insufficient access to funds of investment projects of renewable usage. Income of foreign capital is partially offset by unstable business climate and economic conditions and partially by the absence of appropriate regulatory framework and efficient system of forcing to follow the requirements of legislation.

The next serious barrier is the lack of long-term loans with moderate conditions. Commercial banks are reluctant to provide loans as the return of long-term investments is risky. Besides that, financial institutions do not have any experience in the analysis of financial aspects of renewable energy investments. Foreign long-term loans are expensive because of the high risk perceived by foreign commercial banks.

Also companies have to bear pre-investment costs. The costs of investment project preparation must be incurred before starting its funding with no guarantee of acquiring funds for project implementation. The absence of demonstration projects increases costs connected with their preparations.

Nowadays, the domestic market of native equipment for power production from renewables is very small. Thus high prices of special equipment produced in small amounts due to the absence of sufficient demand have been formed by now. As a result, companies have to invest a huge sum of money in power plant building.

Finally, access to federal funding, which is necessary, considering technical complexity, high-risk level, and duration of renewable project development, has a significant restriction. Important condition for acquiring funds for a renewable project is compliance with degree of localization which means a certain share of domestic equipment and engineering services usage during the implementation of the project. This course is defined by a number of problems such as an insufficiently skilled workforce, quality and quantity of domestic equipment, etc. One of the main difficulties is fulfillment of requirements of production localization in the set timeframe. The equipment for renewables is produced in Russia, but the technology used in production loses in comparison with imported analogs.

3.2   Information Barriers

Insufficient information about technologies and possibilities of their usage inhibits investment. There is no information about already tried technologies, which are applicable for transfer of existing large fossil fuel boilers to the usage of different kinds of renewables.

Insufficient information about financial, social, and ecological benefits and about rate of investment returns in renewable usage leads to GDP growth limits and environmental degradation.

Absence of reliable information about renewable energy supplies increases a risk of investment. Nowadays there are only preliminary estimates of potentially useable renewable energy supplies.

3.3   Institutional Barriers

Insufficient legislative base in the sphere of renewable development support and inefficient system of measures for enforcement of environmental legislation lead to a complication of projects due to increasing bureaucratic procedures. These negative factors do not contribute to the growth of interest to use more environmentally friendly forms of energy such as renewables.

Local authorities do not take an active part in funding of investment projects in renewable development. The problem was caused by tax legislation. According to the Russian tax legislation, the most part of tax revenues goes to the federal budget. A smaller part goes to a region budget. The local budget has nothing from tax revenues. As a result, local authorities have no money to support any investment projects.

4 Renewable Energy Sources' Usage Stimulation: Experience of Different Countries

4.1 Germany

Nowadays, renewables cover about 10% of total energy consumption in Germany with a constantly growing trend. It would be impossible to achieve such a high rate without targeted governmental support, which means [6]:

- Introduction of a guideline aimed at the improvement of overall building energy efficiency
- Adoption of a law considering energy saving which submits strict requirements for construction companies, which now must design and construct buildings in such a way that energy losses from heating and cooling would be minimal
- Implementation of various premium systems for renewable usage
- Current law considering heating plants which rules out that the grid company is obliged to install units for renewable energy production in short terms and also to reward owners of the renewable units for every kilowatt per hour of power supplied
- Introduction of the law considering taxes on electricity (20.5 Euros for megawatt per hour) which provides tax discounts for consumers whose power is supplied by renewable units

4.2 China

In China renewables cover about 17% of total energy consumption. China is the leader on the renewable market. The Chinese leadership is credited by present reforms aimed to encourage renewable development, the essence of which is as follows [7]:

- Introduction of a law that obliges grid companies to purchase all the electricity produced by renewable power plants.

- Creation of conditions for production of equipment for renewable power stations. Nowadays China is the biggest producer of solar panels.
- Large state investments (more than 50 billion dollars) in construction of mini plants that generate electricity from biogas in the countryside.
- State funding of research and development and staff training.
- Regulation of electricity tariffs and introduction of "green tariffs."

4.3 The USA

The USA is also one of the leaders in energy production based on renewables due to the following government support measures [8]:

- Introduction of federal and state-level tax discounts (10% tax discount for investments in solar and geothermal energy, preferential credit against tax, accelerated amortization, etc.)
- Financial payments for new facilities in generating renewable energy
- System of mandatory quotas in individual states
- Introduction of the Act establishing a system of guaranteed price for energy, produced on renewables' stations
- Regeneration of current

5 Proposals on Stimulation of the Introduction of Renewables in Russia

The Russian Federation Government must be extremely interested in renewables' support, as it will allow the exporting of larger volumes of conventional energy sources which are the main instrument of replenishment of the Russian budget. Renewables' development might create conditions for the production of domestic equipment for renewable energy. Renewables will allow reducing the burden on the environment as the amount of energy produced by TPP using coal, fuel oil, and diesel fuel will reduce. Following mechanism of state support is suggested for stimulation of renewable usage:

1. There are many areas in Russia with population isolated from a centralized power supply (some districts of Sakha Republic, Kamchatka, Murmansk region, etc.). Supplying consumers with power in these areas is carried out by imported fuel (coal, fuel oil, and diesel fuel) that leads to the high cost of power production in these districts; that is why the renewable development here can be economically justified (savings on fuel). There were some calculations (on the example of energy complex which consists of diesel power station and photovoltaic power plant in Bagatai town, Sakha Republic) that allowed reducing production costs by

7% [9]. However, high capital costs will not allow making implementation of renewables economically attractive. State support is required to attract private investors. Unlike the existing support scheme based on power supply agreements, the author suggests to introduce premiums for each kilowatt per hour sold, which makes the procedure of state support acquiring more clear and predictable for business.

2. Due to stimulating the renewable competitiveness with conventional resources on domestic market, it is necessary to legally limit grants, subsidies, and tax incentives to hydrocarbon producers (Gazprom, Rosneft, etc.).

3. Temporary cancellation of a rate that considers the degree of localization of the equipment used for renewable energy plants. This will allow applying more modern, reliable imported technologies and equipment in construction right now and also will make domestic manufacturers more competitive in the struggle for Russian consumers.

4. Simplification of procedure for renewable capacity connection to distribution grids. We need to oblige grid companies to carry out connections at their own expense and also to introduce penalties for refusal to connect in the size of renewable producer's losses.

5. Legal confirmation of a long-term contract with the renewable energy supplier for 15 years at least. This will allow minimizing risks for the supplier.

6. Finally, it is only natural that the primary importance is acquired by establishing a proper legislative base. The result should be a design of clear procedure to obtain the budget financing for everyone who wants to invest in renewables and changing tax legislation in favor of local budgets.

6 Conclusion

The suggested measures on stimulation of renewable introduction in Russia will allow making domestic energy market more competitive. State financing will cause an increase of investments in the renewable sector and also will trigger a multiplicative effect with the help of the development of related industries. Renewable "green energy" will allow us to reduce the negative impact on the environment.

References

1. Prediction of long-term socio-economic development of the Russian Federation for the period till 2030. http://government.ru/media/files/41d457592e04b76338b7.pdf
2. Russia's energy strategy until 2030. http://www.minenergo.gov.ru/activity/energoeffektivnost
3. Bezrukyh, P.: Current situation and trends of renewable energy in the world. S.O.K. **8**, 15–20 (2014)
4. The state programme of the Russian Federation energy conservation and increase of energy effectiveness until 2020. http://rusecounion.ru/sites/default/files/energysave_2020.pdf

5. Popel, O.: Renewable energy in the regions of the Russian Federation: problems and prospects. Energosvet. **5**, 22–26 (2011)
6. Gubanov, M.: Features of the German legislation on energy efficiency and renewable energy. Ind. Power. **1**, 54–61 (2013)
7. Chesnokova, S.: China keeps leadership in the development of renewable energy. East Analyst. **3**, 161–164 (2012)
8. Dakalov, M.: Standard and legal regulation the use of renewable energy in the United States: key documents. Bus. Law. **1**, 224–226 (2013)
9. Boyarinov, A.: Encouraging the development of renewable energy in Russia. In: Y. Adamyk (ed.) Innovations in Modern Science. Materials of the VIIth International Winter Symposium 02/27/2015, pp. 8–12. Sputnik+, Moscow (2015)

Energy Efficiency and Environmental Friendliness of Production as Factors of Consumer Value of Goods and Image of the Enterprise

I. V. Ershova, N. V. Dukmasova, and M. A. Prilutskaya

1 Introduction

The sociological polls of the population show that energy efficiency and ecological safety of production become the important component of public consciousness. At the same time, there is a big gap between understanding ecological safety and energy saving problems and perception of environmentally friendly and power effective technologies and products as a consumer benefit. Many authors of the research about consumer value of goods formation share the opinion that the value consists of the competitive benefits and should be measured in terms of money. Some researchers include the social benefits connected with the utilization and ecological safety [1] in the structure of the goods consumer value, but the process of financial measurement and accounting of these benefits, especially for the industry, is not formalized.

The goal of the work described in this chapter is to justify the methodical approach to the assessment of the ecological management system influence on the consumer value of the goods and producer image. More than 300 respondents were interviewed between 2008 and 2012 about their views on environment-friendly goods and energy-saving production, knowledge of eco-labeling, and ecological management systems. The results of the poll are given in the chapter. Their analysis shows that for the Russian buyer, these components of the industrial goods consumer value are insignificant and highly volatile. The share of ecological part in the general consumer value of the goods does not exceed 10%. This should be considered by the industrial enterprises during their price strategy and ecological actions development.

I. V. Ershova · M. A. Prilutskaya (✉)
Department of Industrial Business and Management, Ural Federal University, Yekaterinburg, Russia

N. V. Dukmasova
Department of Environmental Economics, Ural Federal University, Yekaterinburg, Russia

© Springer International Publishing AG, part of Springer Nature 2018
S. Syngellakis, C. Brebbia (eds.), *Challenges and Solutions in the Russian Energy Sector*, Innovation and Discovery in Russian Science and Engineering,
https://doi.org/10.1007/978-3-319-75702-5_25

2 Accounting Features of the Environment Friendliness for the Industrial Production Goods

Depending on the nature of consumption, consumer goods can theoretically be divided into two groups: the first is essential goods (food, clothes, etc.); the second is the technical goods of industrial production. Concerning the first group, the ecological component in comparison with the energy efficiency is the defining characteristic of production and can be estimated quite easily. For example, under eco-friendly products we understand goods grown in personal gardens and farms, so they are mostly products of home production. Speaking about them, the consumer usually notes that this production was done without pesticides, herbicides, and growth factor usage and that there are no preservatives or colorants. It should be noted that many producers often use the word EKO or BIO on their labels, but it does not guarantee ecological purity of the goods. Such marketing mix is used to attract buyers. The consumer buys goods due to the low awareness in this sphere without considering whether the purchase differs from the others or whether this marking has anything to do with the ecological properties of the goods. Often the consumer does not correlate the ecological quality of the products to the sanitary and hygienic norms of packaging, storage, sale, and transportation of the foodstuff defining their safety. Nevertheless, more often buyers pay attention to the ingredients of the product. It means that people started to increase their ecological awareness. These questions are in detail considered in the scientific publications of Heijnen [2].

Difficulties arise when the consumer attempts to estimate the ecological purity and energy efficiency level of the technical goods of industrial production (characteristics of environment friendliness can differ, depending on the production type). The national standard of the Russian Federation GOST R ISO 14040-2010 "Ecological management-Assessment of life cycle-Principles and framework" defines the structure of technical production ecological properties. According to the standard, the good is "eco-friendly" if at each stage of its life cycle it causes the minimum damage to the environment and human health. The life cycle includes several stages: production and processing of raw materials, transportation, production, usage, utilization, or recycling (processing) of goods. The goods can be considered "eco-friendly" and "power efficient" if they are made with the minimal energy and natural resources consumption and can be processed with the minimal environmental load after they were used [3].

In order for the consumer to find and recognize eco-friendly products easily, eco-labeling was created more than 30 years ago. Ecological marking is ratified by the ISO 14020-2000 and GOST R ISO 14020-2011 standards "Environmental labels and declarations-General principles." The eco-labeling indicates the harmlessness of the goods for human health and the environment.

3 Research of the Consumers' Attitude to Eco-friendly and Energy-Efficient Goods

The authors conducted research to identify changes in the idea of the eco-friendly and energy-efficient goods among potential consumers. Within 5 years, consumers of different ages, sexes, and social statuses were surveyed about the ecological and power management and ecological qualities of the goods. More than a total of 300 people took part in the survey.

The result of the survey showed that the number of people who are ready to pay more for eco-friendly goods (made on energy-saving productions) is growing every year, which is confirmed by the non-Russian authors' research [4]. Also, the percentage of money consumers are ready to overpay if the goods really meet all the requirements imposed to eco-friendliness and energy efficiency (Fig. 1) increases.

The uncertainty and subjectivity of the terms "eco-friendly goods" and "energy efficiency of production" influenced the respondents' answers. Generally, the consumers consider that these goods should be safe to use. In Russia, these terms are not ratified at the federal legislative level, and there are no standards of the eco-friendly goods [5].

Based on the results of the survey, it is possible to note that the number of people paying attention to eco-labeling increases (Fig. 2) each year, but does not form the majority.

It is possible to conclude that people want to buy eco-friendly goods, but they do not know how to choose them and that they simply do not know the ecological labels. For the energy efficiency of production, such marking is not even developed.

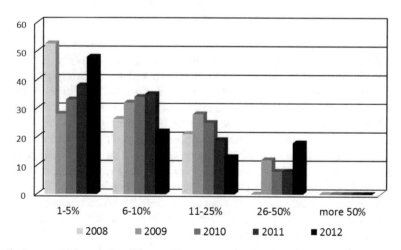

Fig. 1 Answers to the question: "How much are you ready to overpay for eco-friendly and energy-efficient goods made energy-saving technologies?"

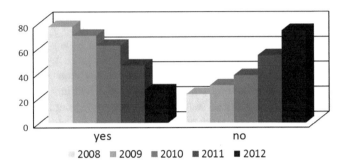

Fig. 2 Share of the consumers paying attention to the eco-labeling while choosing the goods

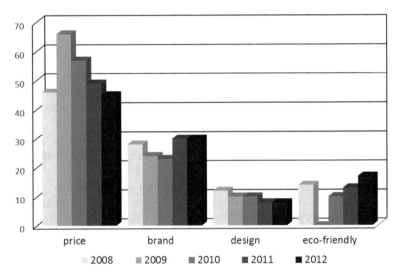

Fig. 3 Consumers' priorities at the choice of goods

Considering that, today, the definitions of terms "eco-friendly goods" and "energy efficiency of production" are not ratified for the consumers, the buyers were asked two additional questions to clarify their position. The first question was connected with the factors influencing the consumer's choice: "What priorities do you follow to make a purchase?" The results of the poll are given in Fig. 3.

The results of the poll show that the priorities of ecological benefits strongly fluctuate depending on the economic situation and are not deciding for the buyer. So in 2008 the price (46.5%) and brand (28%) of the goods were the most important criteria for the consumers and not the ecological component (14%). In 2009 the share of eco-friendliness decreased to 0%; it can be explained by the international economic crisis. In 2010 the situation changed: the share of the price was 57.5%, brand 23.7%, design 10.5%, and eco-friendliness 8.3%. In 2011 the situation changed again: the priority of the price decreased (49%); 30% of the respondents chose a product based on brand and 8% on design; the share of eco-friendliness increased to 13%.

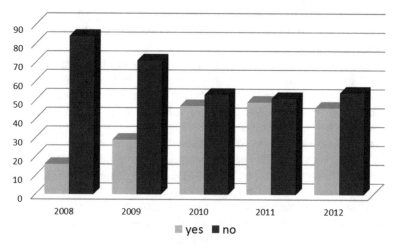

Fig. 4 Consumers' awareness on the existence of the ecological and power management systems

In 2012 economic indicators changed once more, and the priority of eco-friendly goods increased to 16%.

The next poll was designed to determine the consumers' awareness level on the ecological and power management systems at the enterprises. The question was: "Do you know about the existence of the ecological and power management systems at the Russian enterprises?" The results of the opinion poll are given in Fig. 4.

As was assumed, today more and more people can say with confidence that they know about the ecological and power management systems. Thus, in 2008 the share of the respondents who answered "yes" was 16%, and 84% of the respondents answered "no." In 2009 these figures changed: 29% against 71%, respectively. In 2010 47% answered "yes" and 53% said "no" and in 2011 49% and 51%, respectively. The unexpected results were received in 2012: 46% answered "yes" and 54% of the respondents said "no."

4 Results' Interpretation of the Consumers' Opinion Poll

The instability of the consumers' estimates allows us to assume that eco-friendliness of the goods and energy efficiency of their production are not considered as a substantial consumer value. After determining the average price increase for the eco-friendly goods produced with the energy-saving technologies based on the most significant sample frame (the first and second groups, Fig. 1), it is possible to note that the maximum price growth for the eco-friendliness and energy efficiency is in the range from 5% to 7%.

The received results in many aspects are correlated with the large-scale research conducted by Yermolaeva [6] on the Russian and American students' sample list.

The published results prove that along with the high ecological concern, the ecological culture and pro-ecological activity of the Russian youth are developed and formed rather slowly.

Perhaps, the growth of the cost value factors should be searched not in the ecological quality of the goods as this concept for technical production, unlike food, is unstable and subjective, but in the environmental friendliness and energy efficiency of the production technologies [7]. Not incidentally, the second priority for goods after the price is "brand."

The authors assume that the concept "brand" directly correlates with the concept "image of the producer." The eco-friendliness of the production technologies is confirmed by the introduction of the ecological management at the enterprise and availability of the ecological certificate. In Russia the process of the ecological management system implementation began in 2002, mostly at the metallurgical, extracting, and processing enterprises [8]. The mentioned enterprises export their production, and the availability of the ecological management certificate is a necessary condition of the production deliveries to the developed countries' markets. Approximately since 2007 the ecological management has been introduced at the enterprises of other branches and in a service sector. Along with it, the Russian enterprises use the existing standards and laws in the field of energy-saving, which also allows the ratification of the compliance assessment with the energy efficiency modern requirements [9].

According to the authors' previous research, the ecological component of the goods value in the direct deliveries of the industrial production to the foreign markets fluctuates within 5% [10]. An important task is to distribute this practice on the domestic markets, because in accordance with the authors' research and other ecologists [11], the costs of the ecological measures introduction in Russia do not pay off in spite of decrease in ecological payments and penalties.

5 Conclusion

Today, Russian consumers do not correlate the eco-friendliness of goods and energy efficiency of the technologies directly to the consumer value, but the tendency of awareness growth in this area is revealed. The research expects that in the near future, the buyer will become more ecologically competent and will consider such characteristics as power consumption, water-retaining capacity, and impact of goods on health while choosing the product.

The practice of the export enterprises shows that the approximate share of the ecological component in the production price of both consumer and industrial goods is about 5%. This factor should be considered by the enterprises positioning their goods as eco-friendly and made on the basis of the energy-saving technologies. The increase in consumer value should be reflected in the price policy and in the assessment of the ecological and power management introduction results.

Today, the most effective ecological and power management introduction incentives at the Russian enterprises are direct administrative measures from the state and direct influence of the foreign partners demanding the ecological certificate. The conducted research showed that with the correct marketing policy, it is possible to expect traditional ecological and economic effects, along with an increase in prices by 5% and receiving the additional "image" component of the economic effect.

References

1. Shchegolev, V.V.: Methods of the industrial output consumer value assessment. Sci. Tech. Sheets SPBGPU. **3**(99), 68–76 (2010)
2. Heijnen, P.: Informative advertising by an environmental group. J. Econ./Zeitschrift fur Nationalekonomie. **3**, 249–272 (2013)
3. National standard of GOST R ISO 140402010 of the Russian Federation "Ecological management. Assessment of life cycle. Principles and structure". www.gostrf.com/norma_data/58/58831/index.htm
4. Ariwa, E., Okeke, O.J.-P.: Green technology and corporate sustainability in developing economies. In: Proceedings – 6th International Symposium on Parallel Computing in Electrical Engineering, PARELEC 2011, pp. 153–160 (2011)
5. Boyarinov, A.Y., Magaril, E.R.: Improvement of scientific and methodical bases of the production ecological costs compensation economic mechanism formation, UGTU-UPI. Economy Manag. Ser. **5**, 96–106 (2010)
6. Yermolaeva, P.O.: Ecological culture of the Russian and American students. Sociol. Res. **12**, 12–19 (2012)
7. Anufriyev, V.P., Lobanov, V.: We head for energy efficiency. Energostyle. **3**(28), 18–24 (2014)
8. Russian Federal State Statistics Service, www.gks.ru
9. The State Program of the Russian Federation Energy Conservation and Increase of Energy Effectiveness until the Year 2020, http://rusecounion.ru/sites/default/files/energysave_2020.pdf
10. Dukmasova, N.V., Yershov, I.V.: Methodical approaches to definition of the ecological management system introduction economic effect. Messenger URFU. **6**, 34–39 (2013)
11. Vershkov L.V., Groshev, V.L., Gavrilov V.V.: Prevented Environmental Damage Identification Methodology (main editor Chair of the State Committee for Environmental Protection Danilov-Danilyan, V.I), [in Russian]. State Committee for Environmental Protection, Moscow (1999)

Recovering Production and Energy Potential of Hazardous Production Facilities After Natural and Industrial Disasters and Catastrophes

A. M. Platonov, V. A. Larionova, and S. V. Polovnev

1 Introduction

When natural and industrial disasters and catastrophes happen, first of all, there occurs a dramatic transformation of different energetically saturated flows (electric, chemical, nuclear, bacteriological ones, etc.) with an instant loss of flexibility control of their energy in space, in time and as per the territories.

The most dangerous risk situations (RS) for the enterprises that are connected with the control of energetically concentrated flows may occur and quite often do occur at the hazardous industrial facilities (HIF), extremely hazardous industrial facilities (EHIF) and facilities, critically important for the territories (CIF) (i.e. dams, pipelines, water-storage ponds, etc.) [1–6].

It is necessary that businesses, state and society be ready to conduct emergency recovery operations in the shortest possible time and with minimum expense and to recover the production and energy potential of enterprises.

In case of emergency, the life cycles of enterprises being created and operated (Fig. 1) break and continue again along the line 3–4–5–6, describing destruction (3–4) and recovery (4–5–6) processes for the enterprises and their production and energy potentials.

Risk cannot be managed, but the damage from the effects of the RS in the sphere of managing significant volumes of energy at HIF, EHIF and CIF facilities can be considerably reduced due to preparing special design, construction, and reconstruction investment projects being implemented under extreme conditions after the occurrence of natural and industrial disasters and catastrophes [7–10].

A. M. Platonov · V. A. Larionova (✉) · S. V. Polovnev
Academic Department of Economics and Management in Construction and Real Estate Market, Ural Federal University, Yekaterinburg, Russia

© Springer International Publishing AG, part of Springer Nature 2018
S. Syngellakis, C. Brebbia (eds.), *Challenges and Solutions in the Russian Energy Sector*, Innovation and Discovery in Russian Science and Engineering,
https://doi.org/10.1007/978-3-319-75702-5_26

225

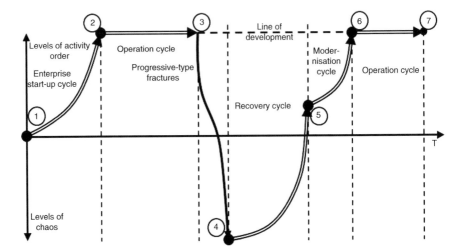

Fig. 1 Change in the life cycles of the enterprises (HIF, EHIF and CIF) before and after natural and industrial disasters and catastrophes, line 1–2–3–6–7 shows the anticipated life cycle of the enterprises; line 1–2–3–4–5–6–7 demonstrates the anticipated life cycle of the enterprises with the account of natural and industrial disasters and catastrophes; and the points 3–4–5–6 represent a "triangle" of conducting emergency recovery operations and repairing the enterprises after natural and industrial disasters and catastrophes

These projects, which must be developed in advance, allow you to reduce the costs of design, construction and recovery operations, to recreate production and energy potential of the enterprises as soon as possible and to return them into the financial-economic and industrial-technological sphere of the country's economy.

2   Modern Realities as the Basis for Production and Distribution of Natural and Industrial Risks, Threats and Hazards

Today, mankind is living, creating and developing in realities that are constantly changing – in the physical reality, and in the reality, created by centuries of human activity, which is in the energetically saturated noosphere [11].

The constantly changing economic and global political realities are, respectively, the third and the fourth realities, also designed by mankind.

Today, the processes of mutual interaction between the aforementioned realities of the external environment take place against the background of the fifth reality implying the dynamic change of mankind's development phases from the ideology of "modernity" to other stages, i.e. "counter-modernity" and "post-modernity" [12].

All the realities described above are in constant motion, interacting, crossing and mutually influencing each other, and in their negative synergy form a united,

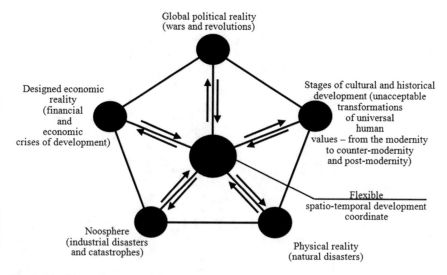

Fig. 2 Flexible spatio-temporal development coordinate in the system of real and critical risks, threats and hazards

temporal and endlessly dynamic "development coordinate" of mankind, ethnicities and states, regions, cities and enterprises (Fig. 2).

Thus, mankind faces not a blissful future, but a future with qualitatively different development risks, more severe than before.

Here, the risks of loss of control over production and energy potential of HIF, EHIF and CIF facilities under the conditions of mutual influence of critical realities remain the highest, and their effects are the most devastating at all stages of transformation of the energy potential of production facilities – from extraction, preparation and utilisation up to energy saving and recycling (Fig. 3).

Here the possibility of integrated influence of the aforementioned realities on the processes of human development and enhancement of negative synergistic devastating effects cannot be doubted.

Only a systematic approach to predicting possible serious disasters and their consequences at HIF, EHIF and CIF facilities (based on statistics and their own experience) can be a response to the given challenges [7–9].

The next step for further development of this approach is the development of respective investment and construction (extreme) projects on recovering the destroyed production and energy potentials, facilities and infrastructure of enterprises [10–15].

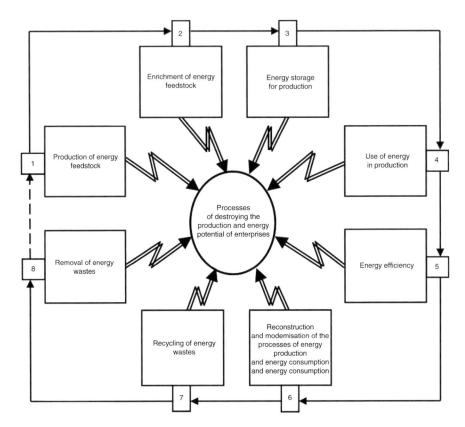

Fig. 3 Energy potential transformation processes as per technological stages of production and utilisation and energy saving, recycling and elimination of energy

3 Setting Up Extreme Projects of Recovering Production and Energy Potential of Enterprises

The practice-oriented algorithm of setting up investment and construction (extreme) projects on recovering production and energy potential of enterprises starts with the determination of the hazard levels of their plants, industrial sites and particular territories depending on volume and power of energy and on the size of possible damages (green, yellow, orange and red levels) and ends with the compilation of the respective maps of hazards (risks).

The conducted analysis allows to define the places of energy concentration that are the most dangerous in terms of emergency occurrence.

The system of hazard (risks) maps on predicting disasters and catastrophes at the potentially hazardous facilities is created on this basis.

Three possible contradictions between the level of threats and the measures being developed can occur when conducting emergency recovery operations at the enterprises and recovering their production and energy potential. In particular, these contradictions can occur between the existing potential of managing the enterprises and the strategy of conducting emergency recovery operations, between the threats of disasters and catastrophes and the level of strategies for conducting emergency recovery operations and repairing the enterprises, as well as between the system of objectives of conducting emergency recovery operations and repairing the enterprises and a set of traditional and extreme investment and construction projects for conducting emergency recovery operations and repairing the enterprises.

On the whole, a practice-oriented design approach to the arrangement and functioning of the extreme recovery system of the enterprises' production and energy potential after emergencies should involve management of content and time of implementation, cost, resources, quality, risks, contracts and interactions among those participating in the implementation of the extreme project under the emergency conditions.

The experience of scheduled reconstruction of the enterprise's facilities shows that the proportion of preparatory period operations reaches 15–30% of its complete cost. It is obvious that the reconstruction of the destroyed facilities and infrastructure of the enterprises after emergencies will require much greater efforts for carrying out preparatory operations. In the absence of statistical data, those works can be estimated within 30–50%.

Success can be achieved by means of timely preparation for carrying out design, construction and recovery operations.

Such kind of preparation involves authorities' and facilities' timely preparation of recovery sets of design documentation on developing extreme projects and establishing appropriate groups and departments for carrying out design, construction and recovery operations, preparation of the enterprises belonging to municipal facilities and construction sector for carrying out recovery operations and creation of the necessary bank of material and technical resources.

The design documentation package for the development of the extreme project on carrying out design, construction and reconstruction operations in conducting emergency recovery operations after disasters and catastrophes at the facilities of the enterprises should involve predictions of possible destruction of buildings, structures, equipment and process lines; possible options for conducting recovery operations at the facilities; resources necessary for carrying out recovery operations; possible schemes to provide power, transportation and means of communication for the recovery operations at the facilities; sources to provide materials, labour, tools and equipment for the recovery operations; volumes and sources of getting the required assistance and the procedure of interaction between the participants of recovery operations; and schemes of notifying managers and systems of managing recovery operations at the facilities.

In addition, the timely prepared set of project documentation for repairing the enterprise should include:

• Explanatory note containing general layouts of the facilities and their compositional justification; space planning and constructive solutions for the main buildings and structures; data on the design capacity of the facilities (number of working shifts, daily schedule, etc.); main technological solutions; solutions on providing the facilities with engineering networks, communications and engineering equipment; and fire prevention measures and engineering and technical measures for Civil Defense and Emergency Situation Management at the facility

- Situational plan, showing building lines; site borders; sanitary protection zones; names of streets and driveways; existing and planned buildings and structures, including the adjacent areas; numbers of buildings (structures); explanation of buildings; and their number of floors
- Technological plans with the layout of large and unique equipment, transportation facilities at the areas and layout of all basic technological equipment
- Utility flow diagrams and existing and planned communication lines

The main requirement for the organisation of design, construction and recovery operations at the facilities after natural and industrial disasters and catastrophes is to provide for the minimum time limits for carrying out such operations, with the account of limited resources.

In general, the organisational and economic mechanism of an extreme investment and construction project is a form of cooperation between the project participants that is fixed in the project materials and in the regulations on conducting emergency recovery operations and repairing the enterprises in order to ensure the implementability of the project within the shortest possible time, to provide for the required level of quality and the possibility to measure the costs and results of each participant involved in the implementation of this project.

Organisational and economic mechanism of implementing nontraditional (extreme) investment and construction projects, when conducting emergency recovery operations at the hazardous industrial facilities, also includes:

- Obligations imposed on the participants of nontraditional (extreme) investment and construction project and taken upon by them for carrying out joint actions on carrying out emergency recovery operations and repairing of the enterprises, as well as guarantees of such obligations and sanctions for their violation
- Emergency procedure for financing the project with the aim of implementing it within the limits of the allocated funds in the shortest possible time and with the specified quality of the works performed
- Special conditions of production turnover and resources turnover among the participants of nontraditional (extreme) investment and construction projects (barter, preferential prices for mutual settlements, provision of commodity loans, compensation-free transfer of fixed assets on a permanent or temporary basis, etc.)
- Necessary synchronisation of the activities of individual participants and timely adjustment of their further actions in order to successfully complete the project as soon as possible
- Measures for the mutual financial, organisational and other support of the participants of the investment and construction project under the emergency conditions and under the conditions of extreme shortage of time (providing temporary

financial assistance, loans, payment delays, etc.), including measures of state support

• Measures of the state (or the third party) compensation of the expenses incurred by the participants of the investment and construction project while conducting emergency recovery operations and repairing the enterprises

For the successful implementation of the measures described above, it is necessary that a change in the engineering legislation be made and the concurrent design of the process of repairing plants, industrial sites and infrastructure, subject to potential threats of being destroyed, be introduced (at the cost of the customer or the state) into the practice of designing HIF, EHIF and CIF facilities.

4 Conclusion

Unlike the traditional management of investment and construction projects, the management of extreme projects on recovering production and energy potential of the enterprises after natural and industrial disasters and catastrophes has its own peculiarities both in terms of the pace of their implementation and in terms of quality and cost of operations.

The given extreme projects are, first of all, characterised by the unpredictability of the time and place of emergency situations and, consequently, by the urgent need for recovery operations.

Therefore, timely developments of extreme recovery projects on recovering the enterprises, destroyed by natural and industrial disasters, are today vitally necessary in order to ensure the necessary level of the society's readiness to deal with risks and threats of destruction of production and energy potential of HIF, EHIF and CIF facilities.

References

1. Beck, U.: Risk society: Towards a new modernity. Progress-Traditsiya, Moscow (2000)
2. Porfiryev, B.N.: Risk of natural and industrial disasters in the world and in Russia. Russia in the surrounding world: 2004. In: Marfenin, N.N. (resp. ed.), Marfenin, N.N, Stepanov, S.A. (gen. eds.) Analytical Yearbook. Modus-K-Aeterna, Moscow (2005)
3. Munich Re: Press Release. 29 Dec 2003, http://pulse.webservis.ru/Science/MunichRe/2003.html
4. Industrial disasters: history and future, http://www.i.stroy.ru
5. Bezzubcev-Kondakov, A.: Why Did this Happen? Industrial Disasters in Russia. Piter, Saint-Petersburg (2010)
6. Osipov, V.I.: Natural disasters at the turn of the twenty-first century. Probl. Saf. Emerg. Situations. 1, 54–79 (2001)
7. Internet industrial disasters (Information about the Most Terrible Industrial Disasters). http://chronicl.chat.ru/tehnogen.htm

8. Shchegolev, V.: Management in emergency Situations: experience, suggestions and prospects. Grazhdanskaya Zaschita. **11**, 40–41 (2000)
9. Arkhipova, N.I., Kulba, V.V.: Management in Emergency Situations, Russian State University for the Humanities, Moscow (2008)
10. Fedoseyev, V.N.: Prevention of emergency situations and conducting emergency recovery operations (managerial aspect). Manag. Russ. Abroad. **6**, 72–79 (2001)
11. Vernadsky, V.I.: Biosphere and Noosphere. AIRIS-PRESS, Moscow (2003)
12. Kormyshev, V.M., Sachkov, I.N.: Introduction into Synergetics. FortDialog-Iset, Ekaterinburg (2013)
13. DeCarlo, Doug: Extreme project management. p.m. Office, Yossey Bass (2005)
14. Egorov, A.N.: Construction Management for Emergency Situations (Theory and Methodology): Monograph. SPbGASU, Saint-Petersburg (2005)
15. Egorov, A.N.: Managing construction and recovery operations in the extremely urgent situations. Econ. Manag. **1**, 87–90 (2006)

Part VI
Personnel for the Power Industry

Individualised Learning Trajectories for the Professional Growth of Managers in the Energy Sector

L. D. Gitelman and A. P. Isaev

1 Introduction

Having more professional managers and developing their professional skills constitute a key competitive edge for any business. The problem is particularly relevant in the energy sector that has to deal with a risk of disastrous emergencies caused by the declining reliability of technological systems.

There are three key industry-specific components of a manager's professionalism in the energy sector:

- The ability to do the job effectively and deliver high performance results in a stable manner in strict compliance with reliability, safety and environmental standards
- Commitment to one's production tasks and functions
- The ability to improve one's knowledge and competencies in order to prepare for the handling of new tasks [1]

Due to an extended investment cycle, the energy sector also requires a professional manager to be able to take forward-looking decisions.

The *professional development* of a manager is the process of acquiring necessary qualities and competencies and personal development skills. Professional development is a complex process that is governed by general patterns but has its own features that determine a specific development path for each manager. For the process to have a direction, scale (corporation, company, unit, worker) and steady dynamics, it needs to be managed. This chapter suggests management tools for professional development during the particularly challenging postcollege period.

L. D. Gitelman · A. P. Isaev (✉)
Department of Energy and Industrial Management Systems, Ural Federal University,
Yekaterinburg, Russia

© Springer International Publishing AG, part of Springer Nature 2018
S. Syngellakis, C. Brebbia (eds.), *Challenges and Solutions in the Russian Energy
Sector*, Innovation and Discovery in Russian Science and Engineering,
https://doi.org/10.1007/978-3-319-75702-5_27

The authors are convinced that creating a friendly environment for self-learning in the company is the most effective method of managing the onsite professional development of managers. The environment should be tailored to the worker's personality, his or her professional development needs and interests.

The individualised learning trajectory is a promising methodological concept for corporate training systems. The instrument is well fitted to be used for customised training of the most sought-after professionals.

2 Research Methodology

The *individualised learning trajectory* is a self-directed learning plan that combines the needs of the company to address production tasks that require new competencies and the worker's own professional development needs.

A number of *indicators* are used to build individualised trajectories in line with the approach:

- Personal qualities of staff members, including their interests and values
- Aptitude for a career
- Professional degrees, including additional training
- Current professional competencies
- Professional plans and career goals of a manager
- Tasks and functions in the current position
- Needs of the organisation to effectively address new tasks

The indicators are used as a basis for creating methodological tools for building individualised learning trajectories. The latter include:

- Personal profile of a manager's professional growth potential
- A roadmap for realising his or her potential
- Assessment of the organisation's need to further the manager's professional development
- A request for the manager's professional development commissioned by the company

The logic behind and the application of the methodological tools of corporate training customisation occur in two opposite directions: by catering to the interests of the worker and the interest of the energy company (Figs. 1 and 2). The individualised learning trajectory reconciles the interests of both parties as it combines two projects that reflect the interests of the worker and those of the energy company. Professional competency is the key unit of analysis in the study of the manager's performance when designing his or her individualised learning trajectory [2].

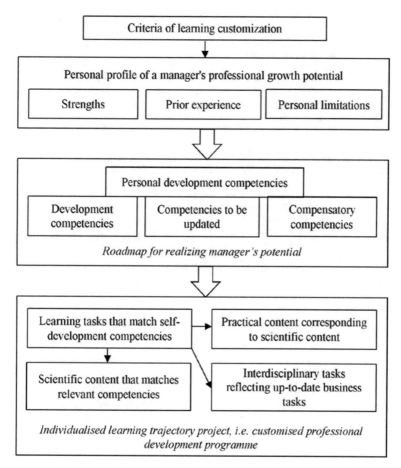

Fig. 1 Basic instruments for corporate training customisation responsive to the manager's personality and interests

3   Research Results

Special psychological diagnostic testing and aptitude assessment methods are used when building *the profile of a manager's professional growth potential*. The profile outlines personal strengths and limitations. Strengths are psychological qualities that fully match the specific nature of the worker's professional functions. Personal limitations are the qualities that disagree with the nature of his or her professional functions, thus hampering their development. The growth potential profile also includes knowledge and competencies that the worker acquired from prior experience and that have not yet been used in the current position but can potentially become driving forces of his or her professional growth. Further diagnosis is needed to identify them.

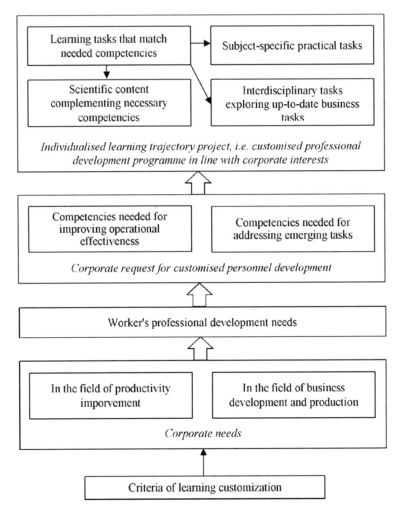

Fig. 2 Basic instruments for corporate training customisation responsive to the energy company's needs

The *roadmap for realising a manager's potential* is a tool for implementing his or her diagnostic profile. It represents the best way to grow professionally by taking into account personality features and adapting prior experience to the current professional activities. It needs to be emphasised that the roadmap for realising one's potential is developed using the competency-based methodology. For example, for each component of the professional growth, potential profile corresponding to *personal development competencies* is designed that incorporates competencies fit for developing, updating and compensatory ones (Fig. 1). The first ones are designed on the premises of making the maximum use of the learner's personal abilities in order to create new professional growth opportunities.

The second ones aim to adapt prior practical experience to solving tasks that match the worker's professional goals. The third ones are designed as modes of professional behaviour that could compensate for the manager's personal limitations (weaknesses).

Managers are able to quickly master self-development competencies because they agree with their personality traits, interests and career plans, provided there is a development programme that is aligned with the roadmap for realising one's potential. An *individualised learning trajectory* could serve as such a programme facilitating the accelerated and quality development of the competencies.

The individualised learning trajectory of a manager takes shape in the course of preparing customised learning tasks and methods of their execution that foster self-development competencies. The tasks are designed based on the subject content and interdisciplinary aspects of the appropriate fields of management and the experience of their application, including the home company. The resulting individualised learning trajectory reflects the manager's personality traits, professional and career goals. Figure 1 shows a step-by-step use of the described toolkit.

As shown above, building an individualised learning trajectory implies the creation of the core elements of self-learning that play a significant role in the manager's development: competencies as the necessary outcome of the learning process aimed at professional development, corresponding learning tasks, web-based scientific content and subject-specific and interdisciplinary learning tasks.

Reciprocally, another training customisation project is crafted that takes into account the interests and needs of the organisation. As a rule, serious businesses are well aware of their issues and development needs and address corresponding challenges. If they are not, personnel training customisation is not only impossible, it is pointless.

A manager's *need to grow professionally* is assessed through the analysis of current efficiency issues and emerging professional tasks. Juxtaposing the needs with the present level of qualification of a manager shows whether he or she delivers on her present production commitments and is capable of dealing with new tasks.

An energy company puts together a request for tailored coaching for a manager by identifying competencies that are required for increasing its productivity and addressing emerging tasks. A training request for one manager will specify one list of competencies, while another request for a different manager will contain another list of competencies, when, for example, the person in question is a candidate for a new key position in the company.

Designing a corporate version of the individualised learning trajectory of a manager is done in a similar way but with the use of a different set of tools as demonstrated in Fig. 2.

The final phase of designing the individualised learning trajectory of a manager is the integration of the two projects. The integrated version of the individualised learning trajectory is created with the direct engagement of the manager. It is a combination of the same elements as listed above: necessary competencies, learning tasks, web-based scientific content and subject-specific and interdisciplinary learning tasks that provide for the manager's professional growth.

The integration of the projects can produce two essentially different results. The first one is when the project built upon the interest of the company and another one built upon the interests of the worker coincide or are close in content. It is the perfect situation because it indicates the employee's motivation and determination to develop his professional skills that are needed by the energy company. Such employees are ready for self-learning, and, more importantly, they are ready to use the results of their self-learning efforts to improve their professional skills and productivity. It is advisable to provide all-round support and encouragement to such workers because investment in them pays off quickly and in spades. Integrating the content of the two projects does not pose much difficulty and produces an optimum learning trajectory. Preparing web-based content and providing methodological support for it are the most labour-intensive parts of creating a personalised continuing education programme based on the integration of the personal and corporate projects for the individualised learning trajectory.

The second possibility is when two projects are radically different content-wise. In this case, it becomes problematic to reconcile production needs and the employee's professional development needs. There are two ways of integrating the learning trajectories that are so different: one is to simply put them together, and the other one is to find a compromise integration solution. The first option essentially means that integration is impossible and results in a decision to implement both learning trajectories successively. The second option implies that the projects could integrate when some parts of each of the two projects are dropped. The resulting trajectory is a loose combination of two projects that, to a certain extent, means a longer learning path and does not fully meet the expectations of both sides involved. The compromise-based integration is not only more challenging to achieve, but it is also fraught with the risk that the manager may not be ready to independently master the suggested trajectory or may not be motivated enough to apply the results of self-learning efforts for increasing his productivity.

4 Conclusion

For managers whose own learning trajectory coincides with or is close to that generated by their company, this instrument becomes an almost perfect means of their professional development and can also be a robust career development strategy [3]. When individualised learning trajectories drawing upon employees' personal professional plans (career goals) disagree considerably with the corporate request, it might be reasonable to first find another position for them and then make an attempt at integrating the two trajectories and creating conditions for self-learning. Of course, the choice of jobs that would match the employee's own learning trajectory is limited, but if the energy company manages to find an appropriate position, it will get an extra chance to improve the quality of its human resources. If no such job is available, then, obviously, it makes no sense for the company to invest in the professional development of the manager.

Individualised learning trajectories are a tool that enables effective planning of corporate training as well as a more effective use of staff resources and the personnel development budget of an energy company. Both functions of the tool make it possible to improve management effectiveness by means of enhancing the professional competence of managers and other staff members and, therefore, increasing the quality of human resources in the company.

Acknowledgement The work was supported by Act 211 Government of the Russian Federation, contract № 02.A03.21.0006.

References

1. Isaev, A.P.: The Basics of Managing Professional Development of Managers in Industrial Companies [in Russian]. Ural Federal University, Yekaterinburg (2010)
2. Baxter, M.: Learning that lasts: integrating learning, development, and performance in college and beyond (review). J. High. Educ. **73**(5), 660–666 (2002)
3. Solberg, V.S., Phelps, L.A., Haakenson, K.A., Durham, J.F., Timmons, J.: The nature and use of individualized learning plans as a career intervention strategy. J. Career Dev. **39**(6), 500–514 (2012)

Visual Brainstorming in Concept Project Development in the Power Industry

O. B. Ryzhuk, L. D. Gitelman, M. V. Kozhevnikov, O. V. Bashorina,
A. Boyarinov, E. A. Buntov, A. V. Kuzmina, O. A. Makarova,
E. S. Pishevskaya, V. V. Polyakova, and V. V. Soloviev

1 Introduction

Visualization as a method of dealing with complex tasks is an area of increasing interest for concept specialists. Experts are inspired by the work of designers and architects, which involves the creation and analysis of visual constructions. Complex tasks tend to share the following characteristics: the unique, challenging nature of the problem, a complete lack of new ideas to address it, tight deadlines, huge amounts of data to be analysed and the participation of many experts from different spheres.

The technological modernization of the power industry and other knowledge-intensive industries is an example of such a task. Innovative projects in a high-tech industry are characterized by the following three features: the long lead time it takes to adopt innovations, which means a long investment lag; secondly, a lot of interdisciplinary links (technologies – economy – environment – human resources); and finally, the overriding priority of technological reliability over business outcomes [1]. All these imply strict requirements imposed on the project rationale.

Visualization [2] significantly saves time and resources spent on problem solving. It also considerably increases the chances of finding an original solution. This study presents the results of project solution developments achieved by applying information visualization.

O. B. Ryzhuk · L. D. Gitelman · M. V. Kozhevnikov (✉) · O. V. Bashorina · A. Boyarinov
E. A. Buntov · A. V. Kuzmina · O. A. Makarova · E. S. Pishevskaya · V. V. Polyakova
V. V. Soloviev
Ural Federal University, Yekaterinburg, Russia
e-mail: m.v.kozhevnikov@urfu.ru

© Springer International Publishing AG, part of Springer Nature 2018
S. Syngellakis, C. Brebbia (eds.), *Challenges and Solutions in the Russian Energy Sector*, Innovation and Discovery in Russian Science and Engineering,
https://doi.org/10.1007/978-3-319-75702-5_28

2 Description of the Visualization Method

Development of a concept project requires a systemic approach, that is, the project should be conceived as a single whole. The integrated picture of a project provides a foundation for the correct solution being formulated and found by different specialists. This integrated picture of the system can be represented by a visual image which is elaborated through brainstorming.

Visualization means graphic modelling of systems, for instance, with the help of schemes, diagrams, mind maps, pictures, and storyboarding. Visualization provides an answer to the question: 'What does the structure of the object look like?'

If we are constructing a visual image of an object, it means that we visualize it in order to illustrate its distinctive associative features which, in their turn, will enable us to reveal and analyse its meaningful and functional qualities. The visual image is an answer to the question: 'What does the object look like?'

Visual brainstorming is a method of solving a problem by considering visualized information. That is, the team involved in the process has to 'switch' between the space of the initial information about the problem and the image of the problem, which allows them to see the problem from different angles and thus discover the solution (Table 1).

The visualization method stimulates the team's collective intelligence more efficiently [3].

- The group's intellectual potential increases when the participants can work with information presented in graphic schemes.
- Visual forms complement verbal forms by providing more opportunities for comparison and construction of mind maps.
- People are more actively involved in the task since they have a general picture to which they can add their own details.
- Graphic materials expand the collective memory, which means that it takes less time to improve the group's productivity.

Let us consider the impact of this method by using an achievement strategy as an example. Suppose there is a desired goal which can be represented by a circle and supplied by the appropriate heading. The achievement process means movement towards the goal through a range of steps involving resources and tools [4, 5].

Table 1 The fact area and the association area

Space of the problem, facts and logic	Describe the problem and the restrictions to its solution	Make a plan and choose the tools	Detailed analysis	Analysis of the solution image. Check that the solution conforms to the restrictions
Space of the image, associations	Visualize the description, reimagine the problem	Visualize the information in accordance with the tools	Visualize the connections and search for options	Specify the solution, search for a more precise image of the solution

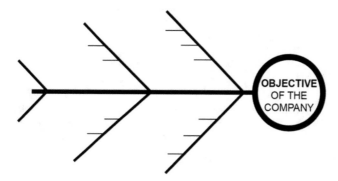

Fig. 1 Ishikawa diagram: classical concept [6]

To show the achievement process, we organize the instruments and resources into a well-ordered structure, reflecting this movement. The question to be asked here is the following: What scheme or diagram can be used to visualize the process of achieving the target? The most suitable type of diagram is the Ishikawa diagram (Fig. 1). Let us analyse the potential of its visual image.

As we can see, the diagram's elements are arranged horizontally, from left to right, which corresponds to positive, successful movement. The target is shown in the circle on the right, by the 'head' of the 'fish' (supplied with an appropriate heading); the main arrow (the 'spine' of the 'fish') is directed at the target and includes additional arrows with smaller 'bones'. The lines are directed towards the target and illustrate the integrated achievement system.

Now the task is to put all the relevant information into this logical structure. If there is no information, then brainstorming begins. What is the heading for the main arrow? Obviously, it should be the strategy. Why are the symmetrical 'bones' necessary? In the case described here, these 'bones' can identify the measures to be taken. They can then be divided into two equal groups, each comprising some more specific steps. Then the 'bones', for tools and resources, are completed (Fig. 2).

This diagram reminds us of a fish skeleton, which explains why it is called 'a fishbone diagram'. Using its characteristics, let us move to the association area and make a deeper analysis of this image. Where does the fish live? What are the peculiarities of its habitat? What prevents the fish from moving forward (for us, this movement corresponds to achieving our goal) and what helps it? What are the threats faced by the fish? What are the opportunities it has? The answers to these questions help us get a more detailed picture of the environment in which we are trying to meet our target (Fig. 3). Let us add some more details to the picture.

Next to the arrows, there is a place for stickers with environmental factors. It is also recommended to supplement this visualization with the quantitative characteristics of each object group. We create, therefore, an integral picture of the system by constructing and analysing the information's visual image. This picture can serve as a basis for making a detailed analysis of the strategy for the achievement of our goal.

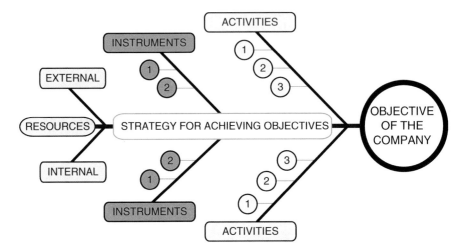

Fig. 2 Ishikawa diagram: information systematization

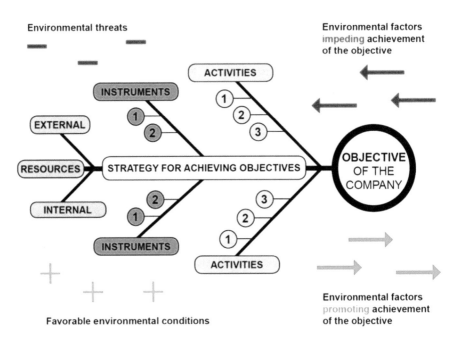

Fig. 3 Ishikawa diagram: environment

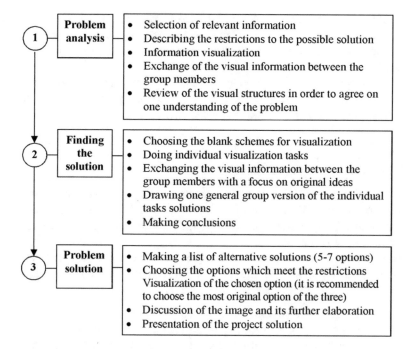

Fig. 4 Stages of practical implementation of the visual brainstorming method

3 Practical Application of the Visualization Method

Thanks to the universal character of visual language, it is possible to grasp the meaning of what your team members have drawn much more easily: this allows the team to spend their energy more efficiently. Visual images bring you to new levels of understanding of the information you are already familiar with.

The skills of information processing are essential for successful implementation of this method (selection of relevant information, representation of the information in the form of structural links, transforming characteristics of the image into the features of the system in question, etc.).

Stages of practical implementation of the visual brainstorming method are demonstrated in Fig. 4.

4 Testing the Method

Practical application of the visual brainstorming method in developing concept projects can be illustrated by the project 'New Leaders for Modernization of Power and Other Hi-Tech Industries'. The project is based on the idea of a partnership between the leading universities and large businesses by creating an integrated technological platform. The project is expected to generate a range of the following commercial products: a system of management training meant for those professionals who have innovative competencies and the tools for stimulating innovation processes in energy companies.

The strategic session involved and included promising managers of energy companies and master's degree students. Summarizing the results of the session, the group had to answer the following questions:

1. Who can this project be sold to?
2. What are the stages of project development?
3. What is the cost of project implementation?

The work was divided into several stages: for the warm-up, the participants carried out the task 'Visual Dictation', which involved making quick sketches of various terms. The goal was to practise the skill of transforming a verbal term into a visual image put on paper.

To encourage the team to answer the questions, the moderators provided blank schemes to be completed with the relevant information. The tasks were done on a paper roll by using felt-tip pens and sticky notes.

Who can this project be sold to? The participants were divided into two teams to construct a customer profile according to the scheme given. The resulting customer profiles were recognizable but also had some unique features. Moreover, the values of the potential customers were revealed [7]. After finishing the work, each group presented their 'customer portraits' and suggested formats for implementing the project in question.

What are the stages of project implementation? To describe the stages, it is essential to understand the nature of the changes the project undergoes by applying the visualization method (see Table 1). Visualization makes it possible to generate innovative solutions, sometimes left unnoticed when following only conventional logic.

The group's collective intelligence can prove to be helpful in finding a more capacious image. Each participant draws an image to reflect the project's features in the best way. Then the participants take turns to present their images explaining why they think these images are suitable. As a result of the presentations and discussion, the group chooses the optimal option. In our case it was the image of an octopus (its characteristics include an ability to multitask, to develop, to regenerate; it also has powerful tools to achieve its targets and a safety system).

Now it is necessary to draw the growth of an octopus from a cell to an adult species, which includes four stages of development: a cell in a test tube, a baby octopus in the fish tank, a young octopus in the ocean and an adult octopus. Each stage of development corresponds to specific growth conditions, environmental factors, and threats and opportunities for project implementation (see Fig. 3). This script can be presented as a table by converting the images into more conventional project characteristics that analysts are familiar with.

Table 2 The formalized result of the visual brainstorming session: the marketing concept of the project

Products	Specialists with innovative leadership competencies	Set of tools for intensification of innovative processes and upgrading the human resources of an energy company	
Markets	Labour	Consulting services	Educational programs
Market segments	Unique specialists	Development consultancies for transformations	Exclusive staff training
Consumers	Government Business entities Universities Research centres Analytical agencies	Business entities Corporate universities Research and educational centres	Gifted bachelor's and master's degree graduates Promising young specialists Managers seeking self-development

In this manner, we get the answer to the second question: a description of the project implementation stages, the threats and opportunities for each stage, the structure of resources and the results of the project implementation expressed in the values of each client group involved.

What are the costs of project implementation? To calculate the costs, it is necessary to specify the functional properties of the project and the resources it requires. Visualization of this question is done in the following stages: project functions; finding the optimal image which would reflect the functions; constructing a more detailed large-sized image; the scheme of resource management and the work of the specific functional parts of the project; and the calculation of costs.

The answer to the third question is the following: the organizational scheme of the project, the implementation stages and the calculation table reflecting the analysis of the cash flow for the project ('New Leaders for Modernization of Power and Other Hi-Tech Industries').

After combining all the visual images, the participants of the strategic session, supervised by the moderators, presented their marketing concept according to the scheme 'products – market segments – consumers' (Table 2).

5 Conclusion

The visualization method described in this chapter is aimed at switching from the usual 'verbal' forms of work with information to working with visual images of the system. This launches the process of devising original solutions for complex problems with a high degree of uncertainty. The method is particularly suitable for teamwork: it increases the efficiency of the collective intelligence along with the objectivity of project solutions.

As experience confirms, in the power industry, this technique can be productive not only in projects developed for specific enterprises but also for complicated tasks shared by the whole sector. For example, when launching programs of energy demand management, described in Ref. [8], the following questions should be addressed:

- Who is the program sponsor: the energy company, consumer or the state?
- What program should be chosen as a pilot?
- What additional financial and legislative initiatives should be adopted?
- How should regional manufacturers of energy-efficient equipment be supported?
- Who is responsible for the results of the program implementation?
- How should demand management programs be connected to the strategies of energy companies and so on?

Thus, the visualization method can be applied as an effective brainstorming tool when developing innovative projects of technological modernization in the power industry. It can also be used for training managers.

Acknowledgement The work was supported by Act 211 Government of the Russian Federation, contract № 02.A03.21.0006.

References

1. Gitelman, L.D., Ratnikov, B.E.: Technical and economic competences as the basis of managers' expertise in electric-power industry, [in Russian]. Energy Market. **12**, 23–27 (2010)
2. Stepanova, T.M., Stepanov, A.V.: Methodological foundation for drawing as a system [in Russian]. Acad. J. Ural Res. Proj. Designing Inst. Russ. Acad. Archit. Constr. Sci. **3**, 87–90 (2012)
3. Brown, S.: The Doodle Revolution. Unlock the Power to Think Differently. Portfolio Hardcover, New York (2014)
4. Archibald, R.D.: Managing High-Technology Programs and Projects. Wiley, New York (2003)
5. Mishin, A.S.: Project Business: Adaptive Model for Russia [in Russian]. Astrel, Moscow (2006)
6. Ishikawa, K.: Guide to Quality Control. JUSE, Tokyo (1968)
7. Clark, T., Osterwalder, A., Pigneur, Y.: Business Model You. A One-Page Method for Reinventing Your Career. Wiley, Somerset (2012)
8. Gitelman, L.D., Ratnikov, B.E., Kozhevnikov, M.V.: Demand-side management for energy in the region, [in Russian]. Economy Reg. **2**, 71–78 (2013)

Author Index

© Springer International Publishing AG, part of Springer Nature 2018
S. Syngellakis, C. Brebbia (eds.), *Challenges and Solutions in the Russian Energy Sector*, Innovation and Discovery in Russian Science and Engineering, https://doi.org/10.1007/978-3-319-75702-5

Subject Index

Printed in the United States
By Bookmasters